THE SIRENS OF MARS

THE SIRENS OF MARS

Searching for Life on Another World

SARAH STEWART JOHNSON

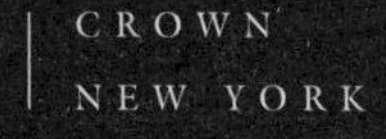

CROWN
NEW YORK

Published in the United States by Crown, an imprint of Random House, a division of Penguin Random House LLC, New York.

CROWN and the Crown colophon are registered trademarks of Penguin Random House LLC.

Title-page image: NASA/JPL-Caltech/Arizona State University
Part-title image: Illustration from *Scientific American,* August 20, 1892, "Professor Pickering's Observation of Mars"
Chapter-opening image: Morphart Creation/Shutterstock

LIBRARY OF CONGRESS CATALOGING-IN-PUBLICATION DATA
Names: Johnson, Sarah Stewart, author.
Title: The sirens of Mars / Sarah Stewart Johnson.
Description: New York : Crown [2020] | Includes bibliographical references and index.
Identifiers: LCCN 2020007280 (print) | LCCN 2020007281 (ebook) | ISBN 9781101904817 (hardcover) | ISBN 9781101904824 (ebook)
Subjects: LCSH: Life on other planets. | Mars (Planet)
Classification: LCC QB641 .J64 2020 (print) | LCC QB641 (ebook) | DDC 576.8/39099923—dc23
LC record available at https://lccn.loc.gov/2020007280
LC ebook record available at https://lccn.loc.gov/2020007281

Printed in the United States of America on acid-free paper

randomhousebooks.com

9 8 7 6 5 4 3 2 1

First Edition

For Emily

Contents

PART 3:
A BOUNDARY IS THAT WHICH IS AN EXTREMITY OF ANYTHING.

Prologue

EVEN WITH THE miracle of modern travel, it takes days for me to reach the edge of the Nullarbor Plain: three or four flights, a quick shower in Perth, then a night or two backtracking east in a rented truck. The two-lane carriageway stretches endlessly through the ghost towns of Australia's old goldfields. Eventually I turn onto dirt roads, onto red rock. When I stop the truck and climb down from the cab, everything is still. Finally, I've arrived at the place where the desert cracks open.

I come here every couple of years, to the ancient terrain of the Yilgarn Craton, some of our world's oldest rocks. Dotted throughout the dark ochre expanse are oval ponds that are as corrosive as battery acid. Yet in these sulfuric waters, against all odds, is the most astonishing array of life. I come to investigate how primitive microbes survive the harsh conditions, how they harvest energy, and what traces they leave behind in the minerals. I come because I'm a planetary scientist and because this is one of the most similar places on Earth to the ancient surface of Mars. I come to the Yilgarn and other wild wastelands—like the McMurdo Dry Valleys in Antarctica, like the Atacama in Chile—to hone my skills at finding life.

Out in the desert, I rise with the dawn. I pull on my tattered field clothes and slather sunscreen across my face. My boots crack with salt as I slip my feet into them. I throw on my hat, with wine corks hung from string around its brim, ready to drive away the flies. I weigh my pack down with equipment and water and head straight onto the flats. I spend my days wading into the sucking mud. In some places the ground parts easily. In others, the enveloping salts are as hard as ice. I note the GPS coordinates, map the terrain, measure the water chemistry, and assess the minerals. Later, back in my laboratory, I will examine each tiny piece of the acid-encrusted world I place into a vial. The sun beats down. The wind whips around me. But I rarely notice. I'm focused, consumed.

At the end of my workdays, after I load up my instruments, I climb up on top of the dusty cab, exhausted. As the sun begins to set, and the sky turns salmon pink, and red dust hangs in the air, it's not hard to imagine that I'm on another planet altogether. Staring off into the silence, I think of all my predecessors, some sitting in deserts just like this, some hoping to signal Mars with giant trenches of fire, others building enormous telescopes in the stock-still air. A boy curled up in the shadow of a Benedictine abbey, longing for his own corner of the unknown to map. A shutterbug from Indiana who developed tens of thousands of blurry images of Mars, hoping one might show something. A French aeronaut who piloted a helium balloon high into the stratosphere, so high that he might asphyxiate, just to get his measurements.

It's a peculiar band I've joined, this pack of Mars scientists, fiercely bound across the generations by the enigma of a neighboring world. One might fairly wonder why we have pinned our hopes for finding life to this red planet. For the last couple of billion years, there has been no rain there. There are no rivers, no lakes, no oceans. Without the driving force of fluid erosion, scars left over millions of years by meteorites are strewn across the surface. Mars has no plate tectonics, no magnetic field, and little protective atmosphere. The terrain is quiet, exposed, and bewilderingly empty.

Yet long ago, before it rusted over, Mars was much more like Earth: smaller, but similar in size and elemental composition. In its early days, Mars was black with igneous rock. Untold piles of lava built the planet's massive volcanic provinces, which bulged with enough basalt to flex the crust. The planet's swollen side cracked opened as Mars cooled, with a fissure so deep that the Grand Canyon could disappear into a side channel. One of the largest mountains in the solar system was formed, towering over an escarpment that itself is nearly as tall as Everest.

Those volcanoes lifted greenhouse gases into the air, wrapping the surface with a blanket of atmosphere. We know from the geologic record that the terrain was warm and wet, at least periodically. Around the time life may have been getting started here—conceivably in volcanic pools, in Darwin's "warm little ponds"—water was present on Mars, pregnant with possibility. In fact, there may have been enough water to fill a northern ocean, still and deep, with a seafloor as smooth and flat as the abyssal plains of the great Pacific.

Then, between three and a half and four billion years ago, our planetary paths diverged, and Mars was laid bare. Almost all of the atmosphere disappeared, and so did the water. The planet slipped into a deep freeze, colder than the cold of Antarctica, leaving Mars the hyper-arid, frozen desert we know today, bathed in high-energy solar and cosmic radiation. Now a dust the consistency of red flour coats the surface, lofted by dust devils into the impossibly thin air.

Yet life, we have learned, is stunningly resilient. It can adapt, it can wedge into a crevasse, it can hang on against all odds, and it can reveal itself in unlikely ways. Traces of biology hide in the most unexpected locations. It's why I roam the terrain at the edge of the world, hunting for the subtlest fingerprints of life, learning how to look.

In the far reaches of Australia, there's one particular lake that stands apart from the others, amid the rocks and dunes, past the Rabbit Proof Fence and Jilbadji Nature Reserve, past the derelict aerodrome. The surface is stippled with halite, a form of table salt that looks like freshly fallen snow. In the right place, with a good grip, you can pull out a crystal of gypsum, severed like a shark's tooth from the jaw of the earth. The spear-tipped blades are as large as your hand. When you rinse away the red mud and hold it to the light, it flashes in the sun like a gemstone. Under a microscope, you can see the tiniest of pockets within it: glinting drops of lake water, sheathed in mineral hideaways. Life caught in a crystalline dagger.

THESE PRISMATIC INCLUSIONS are just one of the many features we want to look for on Mars. We are seeking places where secrets are held, where traces of life might be preserved and protected. For over fifty years, we've been exploring Mars with telescopes, flyby missions, orbiters, landers, and rovers. We've scoured the surface for current life as well as indications of past life, for possibilities and actualities. The wild strangeness of the planet, with its tawny air and relentless red deserts, calls to us: With each mission, we grapple to understand a world that's at once recognizable yet at the same time indescribably foreign. We return again and again, and the mysteries deepen.

In the process, we've built an entire field of science around something we can barely see in the night. Four hundred years ago, Mars was still a blaze of light, no more than an idea. The earliest telescopes showed it about as large as a pea held at arm's length, and even more-modern telescopes gave us little to go on. We had no idea what the surface looked like or what it was made of, if there were mountains or valleys. We had only the crudest of maps. We didn't know if there were clouds or what color the sky was. We started from almost nothing. We've gone careening down blind alleys and taken countless wrong turns, yet somehow, miraculously, the passion, ingenuity, and persistence we have brought to the enterprise have moved us toward a truer understanding of another world.

In this way, the story of Mars is also a story about Earth: how we've sought another stirring of life in the universe, and what that search has come to mean. Mars has been our mirror, our foil, a telltale reflection of what has been deepest in our hearts. We have seen in Mars a utopia. A wilderness. A sanctuary. An oracle. With so few landmarks, guideposts, or constraints, all is possible; without data that could be used to cabin our inquiry or limit our imagination,

Mars has been a blank canvas. And tenderly, our human seeking has rushed to fill it.

As a result, Mars has a human history inscribed upon its surface, even though no human has ever touched it. This book offers an account of our exploration of Mars since the dawn of the Space Age, a relatively recent human effort that has revealed the planet's extraordinary natural history. Most of the explorers in these pages—the modern scientists and the people from centuries past who inspired them—came to Mars seeking connection to something larger than themselves, some piece of evidence, some breakthrough observation that would show life could exist there. And they didn't just seek it, they longed for it—I long for it—knowing that even the smallest glimpse of some greater, deeper, other realm might change everything. This is what sets Mars exploration apart: The quest to bring this distant world into focus, pursued over generations and on the frontiers of technological innovation, has always been about more than scientific knowledge. It has been an almost existential endeavor to confront our own limitations, to learn what life really is, and ultimately to defy our own isolation in the universe.

PART 1

A Point Is That Which Has No Part.

—EUCLID'S ELEMENTS

CHAPTER 1

Into the Silent Sea

IN JULY OF 1965, as a tiny octagonal spacecraft swooped across the Martian surface, my father, who had just turned eighteen, was standing tall on a humid, hardwood-forested hill in Appalachia. There on the edge of Viper, Kentucky—below a hundred kilometers of nitrogen and oxygen, under the Kármán line, the exosphere, and the Van Allen belt, beneath the great, vast vacuum of space—a small natural-gas company had sent a bulldozer up a holler and had set about carving out a flat spot for drilling. On the days my father managed to drive the old jeep through the creek bed without flooding the engine, he joined an overalls-clad, illiterate crew in digging ditches and laying pipe, occasionally carrying the casing for the drill head. He'd hoped to spend the summer as a fledgling assistant to the company geologist, but within two weeks, every available worker had been sent to the hillside.

The news about the world's first Mars mission, *Mariner 4,* came by way of *The Courier-Journal,* the newspaper out of Louisville. It arrived on a truck that twisted along the deeply gouged mountain roads, passed the coal camps, passed Hazard High School, and made its way into the small downtown, which was bound like a bobby pin by the North Fork of the Kentucky River.

That morning, my grandfather had picked up the newspaper from Fouts Drug. He'd tucked it under his arm on his way to work at the health department. As a medical technician, he inspected the Cold War–era bomb shelters that dotted the mountain ridges to make sure the food stocks were safe and drew blood to test for syphilis before young couples got married. He took pride in the fact that everyone in town called him "Doc." He wasn't a doctor, but he did give penicillin shots throughout the hills of eastern Kentucky: down in Gilly, up in Typo, in Slemp and Scuddy, in Happy, Yeaddiss, and Busy. When my grandmother wasn't giving perms, she would help out. She liked running the X-ray machine.

It was still muggy later that evening as my grandfather meandered up Broadway—a street that was anything but broad, a single paved lane that fell steeply into backyards teeming with kudzu. He walked into a house that hung like a bat to the side of the ravine, leaving *The Courier-Journal* in the attic bedroom, which was spacious now that four of the six kids had left home. His lanky, wide-eared child, his youngest son, would also leave at the end of the summer, heading two hours west across the steep forested slopes to attend Berea College. My grandfather put the paper on the quilt where my father was sure to find it, next to his *Popular Science* magazine, right beneath a poster of the pockmarked moon.

My father had been spellbound by the idea of the mission, NASA's chance to photograph the planet most similar to Earth. As the mountain town rotated into darkness that Wednesday, my father climbed the steps, aching and exhausted, and he saw the headline. Above the fold, between a picture of Willie Mays and an article on Vietnam, was what he'd been waiting for: MANKIND, THROUGH MARINER, REACHING FOR MARS TODAY. He smiled and fell into bed as he read. "Today the fingertip of mankind reaches out 134 million miles to Mars, almost touching the only other body in the solar system widely suspected of harboring life . . ."

On the other side of the country, in a canyon north of Pasadena, an eager crowd had gathered on the campus of NASA's Jet Propulsion Laboratory. Inside JPL's von Kármán auditorium, intertwined cables, thick and vaguely subterranean, unfurled from a cluster of television cameras and snaked across the floor to the vans outside. Radio from all over the world was hooked in by relay, and the Brits were poised to broadcast a live television feed, having leased a full two minutes of time from the "Early Bird" satellite. There were thirty-seven phones in varying states of use: thirty-six within the press bank, and one sitting atop a desk as part of a small fake office where the TV broadcasters could be filmed.

From floor to ceiling, dominating one side of the great room, was a full-scale spacecraft, one of the flight-ready spares that had been used for temperature-control testing. It had the same octagonal magnesium frame as *Mariner 4*, the same 260 kilograms of hardware and instrumentation. There were 138,000 parts in all: aluminum tubes, attitude-control jets, pyro end cabling. The solar panels, including flaps at the end, stretched seven meters. Coated with sapphire glass, glistening in the beams of the television lights, they looked like the wings of a jeweled pterodactyl.

Much depended on this craft. In a scene that played out repeatedly over the course of the twentieth century, a Soviet spacecraft was approaching Mars at the same time. It had launched from the Baikonur Cosmodrome just two days after *Mariner 4*. It had reached Mars, but, much to NASA's delight, it wouldn't be returning any data. Halfway there, irregular updates had started coming from its communications systems, and then the transmitter died. It was now no more than "the voiceless 'Russian spy,'" "The 'Dead' Soviet Mars missile." At long last, the United States had a chance to pull ahead in the Space Race.

There was only one hurdle standing in the way of American triumph: *Mariner 4* had to aim and actuate the camera and successfully transmit its images back to Earth. This was no easy feat. Mars was so far from the sun that the mission had only 310 watts of usable power, the equivalent of a couple of lightbulbs. The power available to send the data stream would be a mere ten watts to start, which would dissipate to a tenth of a billionth of a billionth of a watt by the time it was captured in the great dishes of the Deep Space Network, the newly built antennas on the outskirts of Johannesburg and Canberra, and deep in the Mojave Desert. And even if the data arrived, there were worries. What if the pictures snapped a bit too early, or a bit too late? What if the spacecraft inadvertently twisted away from the planet at just the wrong moment? What if the camera failed to shut off, recording over the photographs of Mars with pointless photographs of empty space?

The Soviets had been trying to reach Mars for five years. In space exploration as in all things, they were a formidable adversary. In 1960, their first pair of missions had coincided with Premier Nikita Khrushchev's visit to the United Nations General Assembly in New York. He'd commissioned models of the Mars probes and brought them along to show the world. Less than two months earlier, his lead rocket engineer had launched into space the first sentient beings that returned safely to Earth: two dogs, a gray rabbit, forty mice, two rats, and several flies.

But the Soviets were not so lucky this time. As the delegates assembled in New York, the first rocket to Mars failed, climbing just 120 kilometers before falling back to Earth and crashing in eastern Siberia. Then the second rocket failed: A cryogenic leak had frozen the kerosene fuel in the engine inlet. Khrushchev had been relying on another splendid performance from his ambitious young space program and was furious as he paced the halls of the U.N. Before the plenary meeting came to a close, he supposedly went so far as to pull off his shoe, enraged, and brandish it angrily at another country's delegate.

The Soviets tried again with a trio of missions in 1962. The first ruptured in orbit, fanning out debris that was detected by a U.S. radar installation in Alaska. It was nine days into the Cuban Missile Crisis, and the wreckage was momentarily feared by Air Defense Command to be the start of a Soviet nuclear attack. The third also exploded, the main hull of the booster reentering the atmosphere on Christmas Day, followed a month later by the payload. The second, however, traveled 100 million kilometers away from Earth and went on to make the first flyby of Mars—though it was a mute witness to the event, as its transmitter failed, the same thing that happened two years later.

The Soviets kept their defeats to themselves and trumpeted their successes—which were numerous enough to show that they had a decided lead over the Americans. They had reached practically every milestone in the Space Race: the first artificial satellite, the first animal in space, the first man, the first woman. They'd intentionally crashed a spacecraft into the moon and taken the first pictures of its far side, and they were now poised to claim the first spacewalk.

The United States, by contrast, had successfully completed only one planetary mission, *Mariner 2* to Venus. Worse, the Venus mission, the "Mission of Seven Miracles," had barely worked. It was a wonder that it had managed to collect any data at all, flying by the seat of its pants, "limping on one solar panel and heated to within an inch of its life."

And getting to Venus was easier than getting to Mars. To reach the Red Planet, the spacecraft's systems had to stay alive for an extra hundred days, and the data had to be transmitted twice as far. Transistors were new and bulky, and the microchip had just been invented. The computing power of the whole spacecraft was no better than that of a pocket calculator, yet the spacecraft had to rely on a never-before-tested star tracker to point the way. For the first time in history, a NASA probe was drifting into the darkness, traveling away from everything bright in the night—the Earth, the moon, the sun. Just like Coleridge's ancient mariner, it was poised to be "the first that ever burst/Into that silent sea."

Originally, there were to be two *Mariner* missions to Mars: identical-twin spacecraft, nicknamed the "flying windmills." *Mariners 3* and *4* were supposed to zoom by the planet just weeks apart. But the plan went awry when *Mariner 3* was lost within minutes of launch. The rocket, an Atlas-Agena, had performed beautifully, but it soon became apparent that something was amiss. The data coming back from Johannesburg indicated that the spacecraft wasn't on the expected trajectory. The nose fairing, which was designed to protect *Mariner 3* from the crushing force of the launch, hadn't properly detached from the probe. For nine hours, the operations team fought desperately to find some way to rip it off. They tried everything, including firing the spacecraft's motor, but the batteries finally gave way, and *Mariner 3* drifted into a derelict orbit around the sun.

Mars and Earth align on the same side of the sun only once every twenty-six months, so the team had just a few weeks to engineer a solution before Mars moved out of range. The material for the nose fairing had been fabricated and tested under the dense atmospheric pressure of Earth, and they realized that its honeycombed fiberglass skin had started to pop like popcorn in the vacuum of space, expanding enough to wedge the nose fairing tight. The engineers worked around the clock, eventually recognizing that the honeycombed design could be salvaged by poking tiny holes in each cell to equalize the pressure. Just twenty-three days after *Mariner 3* had failed, *Mariner 4* was, improbably enough, ready to launch. It sat on Pad 37 the night before takeoff, shining under the brilliant spotlights. When morning came, it roared forth, lifting off Cape Kennedy on the wings of an Atlas booster. When the rocket released the spacecraft, the nose shield popped right off, just as it was designed to do.

AS *MARINER 4* left Earth's shadow, it began to roll through space. Its first task was to locate distant Canopus, the reference star its onboard sensor would use for navigation. The onboard sensor locked on to Alpha Cephei, far to the north, then moved southward to Regulus, then Zeta Puppis, then an unnamed cluster of three stars. It finally found Canopus, but soon it locked on to a stray light pattern. Then Regulus again. The sensor lost its way no less than forty times throughout its long cruise, as small dust particles and flecks of paint, shining fiercely in the sunlight, kept getting dislodged and falling into the sensor's field of view. The spacecraft stumbled the entire way to Mars, but it survived the 523-million-kilometer journey, circling halfway around the sun. NASA was only seven years old, and this quarter ton of technology was already being heralded as America's greatest achievement in space.

Inside the JPL Space Flight Operations Facility, among the dark-slacked, short-sleeved men with narrow dark ties clipped to the fronts of their shirts, was Bob Leighton, the head of the *Mariner 4* imaging team. That there was an imaging team at all was itself an unlikely development. The prevailing attitude at the time was that photographs were trivial. "Pictures, that's not science. That's just public information," recalled a project manager for NASA's unmanned *Ranger* missions to the moon.

But Leighton had a deep passion for photography. He knew how to render an image, how to bear witness to a pattern of light. He knew what it meant to really *see* something. He'd grown up poor in California—raised by a single mother, who'd made her living as a maid in a Los Angeles hotel. Not long after Leighton graduated, a high school photography teacher found him a job at a photography lab for a Hollywood advertising firm. He could very well have ended up a professional photographer, had his penchant for neatness and efficiency not gotten the best of him. In 1939, he threw out what he thought was a stack of scrap paper, only to discover that it was a client's underexposed blue-light negatives of an elegant steamship sailing under the Golden Gate Bridge. He got his walking papers and returned to L.A. City College. He transferred into the physics department at Caltech his junior year, then never left, joining the faculty in 1949.

Leighton had spent the decade before *Mariner 4* working at the Mount Wilson Observatory. On just a shoestring budget, he'd built an image-stabilization device, a "guider," and taken photos of the planets. He was supposed to be observing stars and galaxies, charting redshifts, making fundamental discoveries about astrophysics—not watching the planets, which were considered too small, cold, and close to reveal much of anything about the nature of the universe. But he couldn't resist catching glimpses now and again. When no one else was around, mostly on holidays like Thanksgiving and Christmas Eve, he'd sneak peeks through the 1.5-meter telescope.

One of Leighton's students had recently been hired by JPL. He told his colleagues about how his former mentor had been able to remove some of the atmospheric turbulence and make movies up at Mount Wilson, beautiful movies, for the first time in color, of a rotating Mars. When an amateurish TV experiment was finally suggested for the mission, the student pleaded with his old professor: "Bob, as a duty to our community, you've got to make a proposal on the TV experiment to Mars. This other one is terrible."

Leighton acquiesced, and within a matter of months, he designed a gizmo with a slow-scan television camera capable of taking twenty-one photographs of Mars over the course of twenty-five minutes. The hope was to collect images of Elysium, then Trivium Charontis, then sweep across Zephyria and into Mare Cimmerium, catching a glimpse of the desert Electis before petering out to the south of Aonius Sinus. For each image, the camera's shutter would click open for a fifth of a second. The pictures would be recorded on a ribbon of magnetic tape, then radioed back to Earth via *Mariner 4*'s high-gain antenna. In a sense, it would be the world's first digital camera.

For decades, the best telescopic observations, on the best of days, had only brought Mars a few times closer than the moon appears to the naked eye. The pictures from Leighton's system on *Mariner 4* would be incalculably better, resolving features as small as 3.2 kilometers across. They would give the world a firsthand view of Mars. Like Martin Luther insisting on a direct relationship with God, the imaging eliminated the need for an interpreter—or "scientific priest," in the words of the youngest member of Leighton's team. The pictures could be spread all over the world for everyone to see, in a language everyone could understand.

But Leighton knew that photographing Mars was more than just an opportunity to share the mission's findings directly with the public. It would be the culmination of a long quest to *see* an entity we'd never encountered, an entity that had puzzled us for centuries. What *was* Mars exactly?

As INCONCEIVABLE AS it sounds, Mars wasn't always understood to be a place. To be sure, the ancients knew there was something intriguing about Mars. The Mesopotamians noticed that it followed a strange loop in the night sky, drifting separately from the "fixed" stars. Everything in the immense night moved together, everything except five little wanderers. Of those, only one appeared as a blazing red lamp. It wasn't only the planet's distinctive color that made it perplexing but also its motion. Mars drifted eastward, night after night, in relation to the other stars, but for about ten weeks every couple of years, it suddenly turned and backpedaled against the zodiac, wandering west for sixty to eighty days before resuming its normal course. In effect, it traced out an elongated loop. Sometimes the size of the loop was smaller, sometimes larger. From this, Plato concluded that the planets had souls, for what could these retrograde acts be, he reasoned, if not expressions of free will?

It wasn't until Galileo looked through a spyglass from a columned terrace in Padua that Mars began its transformation from a glint of light into a world. Within a short stretch of weeks, it became clear to him not only where Mars was in relation to other celestial objects but also *what* it was. Galileo constructed his perspicillum, or telescope, with his own hands. He mounted it on a stand, and because of its tiny field of view, he had to be utterly still, barely breathing, hoping the falling evening temperatures wouldn't cause the glass to mist over. But through its tiny aperture, he determined Mars to be a spherical body illuminated by the sun.

As to whether this body was like the Earth, Galileo was not sure. He hedged in a letter he sent in 1612: "If we could believe with any probability that there were living beings and vegetables on the moon or any planet, different not only from terrestrial ones but remote from our wildest imaginings, I should for my part neither affirm it nor deny it, but should leave the decision to wiser men than I."

The quest to determine the nature of Mars by seeing it better would continue for centuries. When Galileo first looked through his rudimentary telescope, Mars appeared only the size of a poppy seed. But soon concave lenses gave way to convex ones, inverting the image—a minor inconvenience for a much larger field of view. Focal lengths began to grow, and new lenses reduced the optical distortions that Galileo had tried to mitigate with a cardboard washer. In 1659, the Dutch astronomer Christiaan Huygens stuck a telescope out the attic window of his father's large house in The Hague and sketched the first map of Mars. Within a circle, he drew a v-shaped scribble of ink to denote a dark splotch on the planet's surface. In the nights to come, he watched the shadowy blob disappear and reappear. It looked vaguely like an hourglass. It would later be called the Hourglass Sea, the first known surface feature on the planet. After mounting a tiny measuring device to the eyepiece of his rig, he estimated rather accurately that Mars was about 60 percent the size of Earth.

Much to Huygens's surprise, his observations also revealed that the length of the Martian day was roughly twenty-four hours, practically the same as Earth's. These and other similarities between the two planets fueled speculation about Mars's inhabitants. Huygens assumed the presence of intelligent beings, arguing that the planets could not be viewed as "nothing but vast Deserts, lifeless and inanimate Stocks and Stones," for doing so would "sink [the planets] below the Earth in Beauty and Dignity—a thing that no reason will permit." He went so far as to speculate about extraplanetary mathematics, envisioning tables of sines and logarithms, since "there's no reason but the old one, of our being better than all the World, to hinder them from being as happy in their Discoveries, and as ingenious in their Inventions as we ourselves are."

MEANWHILE, AT CAMBRIDGE University, Isaac Newton was developing the basic optics of a new telescope: the reflector. Lenses at the time could not bring red and blue light to the same focus, resulting in a haze of color surrounding bright objects. So instead of a lens, he designed a prototype that collected light by way of a curved metal mirror: six parts copper, two parts tin. It was a tiny instrument, sixteen centimeters long, yielding a magnification of only thirty-five times. Within a century, however, Newton's design would have sweeping implications, enabling unprecedented enlargements of celestial objects. In 1773, William Herschel, a German-born musician living in Bath, England, began experimenting with the construction of mirrors in his spare time. In collaboration with his brilliant younger sister, Caroline, who went on to discover scores of comets and nebulae, he ended up building dozens of reflecting telescopes. From his south-facing garden, he was the first to detect the faint light of Uranus. He also trained his beautifully handcrafted instruments on Mars, revealing that the planet had white polar caps, "clouds and vapors floating in the atmosphere," and seasonal cycles similar to those of the Earth. In 1784, he gave an address to the Royal Society that cast Mars as a kind of copy of the Earth, noting that its inhabitants "probably enjoy a situation in many respects similar to our own." As the observations about the planet continued to roll in, they all fit with the idea of Mars as another Earth, a planet with its own oceans to sail and lands to walk, a place we could recognize, relate to, and imagine.

The idea that Mars was like our planet only drove the quest to see it better. Newton's reflector telescope was soon surpassed, as the speculum metal used for mirrors in large reflecting telescopes was quick to tarnish, and the polishing process often distorted the curve of the mirror. Instead, refracting telescopes, like Huygens's, using two lenses instead of two mirrors, came back into vogue in the nineteenth century. These gradually grew and grew, with lenses so large that gravity began to cause the glass to collapse in on itself. Important discoveries were made, seasonal changes were tracked, moons were discovered. Telescopes, these magical instruments, transported us and let us see what we had never seen before. For hundreds of years, they were our only way to understand Mars.

NOT LONG AFTER I graduated from college, I convinced my father to come with me on a trip into thin desert air. He got on an airplane with me, something he'd only done a few times in his life, and we flew from Kentucky to Atlanta and then on to Tucson. We rented a car and drove deep into Arizona's Old West country, to a hill over a kilometer above the San Pedro River Valley. There was a little hotel, which closed several years ago, at the Vega-Bray Observatory.

We checked in, then went to examine the telescopes. There was a forty-six-centimeter reflector beneath a roll-off roof, a couple of thirty-centimeter Meades, and a half-meter Maksutov-Cassegrain. We zeroed in on a twenty-centimeter Newtonian scope optimized for planetary viewing, which we moved to the observing deck just after dark. There was no tracking system or computer on it, just a sighting mechanism, which was all we needed. My father knew everything about the night sky.

He had spent a good deal of my childhood in the backyard with sky maps from *Astronomy* magazine tucked beneath his elbow. As much as my father would have loved training for a career in geology or astronomy, he'd needed a job to make ends meet, and he'd found one working with the state health department, just like my grandfather had. I'd seen the night sky many times through his oversized binoculars, which invariably wobbled in my hands, even though he always tried to hold them steady for me.

By this point in my life, I'd been to places like Lick Observatory and Mount Wilson Observatory. I'd spent my summers interning at NASA, and I'd visited the giant domes. I'd seen the data that state-of-the-art telescopes could collect flickering on computer screens. But there was something different about seeing the sky through a medium-range telescope at Vega-Bray.

That night in the desert, I sensed for the first time what Galileo and other early astronomers must have felt, something that's been lost in the age of computers. Planetary science used to be an amateur enterprise. Before the dawn of the Space Age, every single practitioner had a direct relationship with the night sky. They were awake when others slept, alone with their science and their thoughts, enveloped by the vast physical world. To point the barrel of a telescope at a tiny dot in the sky and then see it as a world, that dot, that very one right there! Of all the thousand pinpricks of lights, that one is different. That one's threaded by rings. That one has tiny moons, suspended like marbles. That little alabaster hat is a polar cap. That one is a *world*.

Standing there in the cold night air with my father, the telescope at my eye, I felt connected not only to Mars but to Galileo, to Huygens, to Newton, to Herschel. You can't see something like that and not yearn to see it better. As I squinted, making adjustment after adjustment to the dials of the telescope, all I wanted was to fly up to it. Or at least keep the image still. I cursed the atmosphere that wouldn't allow it—the same sky that keeps us alive, brings us rains, and softens our shadows. Even the rarefied air of Arizona tremored and swirled and maddeningly made heavens flicker away. Caught in the grasp of longing and frustration, I could understand why, by the twentieth century, we had to leave our own planet behind.

TWO HUNDRED AND thirty-one days after *Mariner 4* had launched, on the night of July 15, 1965, the tiny levers of the telex machine at JPL began ferociously clicking. Leighton must have felt a surge of emotion: The *Mariner 4* pictures would be the first ever close-up images of anything beyond the moon, as the mission to Venus hadn't taken any pictures. Leighton poignantly recognized the difference between knowing something about a place and actually seeing it, and so did his imaging team. As Bruce Murray, then only a postdoc, realized, "Looking at a planet for the first time . . . that's not an experience people are likely to have very often in the history of the human race."

The data packets were being flung from Mars to Earth, captured in the huge bowl of the tracking station at Goldstone Space Communications Complex in the Mojave Desert and transmitted across California via teletype to JPL's Voyager Telecommunications section. To Leighton, it seemed that the bits of the picture were like pearls, strung kilometers apart on a string from Earth to Mars. The data rate was only eight and a third bits per second, so it would take eight hours for the first image to be fully transferred. Eight hours of nail-biting, eight hours of pure suspense.

The day before, as *Mariner 4* was approaching Mars, the operations team had decided to relay a command, DC-25, with an updated stretch of code to initiate a platform-scanning action, which would identify the planet, followed by a second command, DC-26, which would ensure the camera stopped and didn't record over the images. The data received before the code was sent suggested that the tape recorder had started and stopped, but there had been some anomalous errors. The tape recorder was also a flight spare, swapped in at the last minute because of a technical problem with the original. It would still be hours, possibly days, before the computers could assemble a real photograph, and now some were second-guessing whether the commands should have been sent, whether they might somehow confuse the computer.

Dick Grumm couldn't stand the wait. He was in charge of the tape recorder, and he and a few other engineers began brainstorming ways they might check the data. It became a contest of sorts, and the winning idea was to print the single-stream data—groups of digits, indicating the brightness of each pixel—onto a reel of ticker tape. As the engineers began snipping the tape into strips and pinning them to the walls, Grumm popped over to a local Pasadena art store in search of six shades of gray chalk, one for each bit of the six-bit image. He ended up with a pack of Rembrandt pastels. "Chalk" was for schools, not artists, he discovered, and anyway, chalk wasn't made in six shades of gray.

Upon his return, a massive paint-by-numbers artwork had been assembled and was now ready to be filled in. The *Mariner 4* image that would eventually be shared with the public was black and white, but Dick's ticker-tape interpretation with colors corresponding to the brightness scale came to life with pastels ranging from light-yellow ochre to burnt umber to Indian red. He had tried a purple color scheme, then a green one. But the red one seemed to best mimic the gray scale. It just happened to mimic the colors of Mars too.

When a jumpy public-relations team got word of what was happening, they went to find Grumm immediately. They didn't want the restless press to seize upon some messy makeshift picture instead of an actual image of the Martian surface. Grumm refused to stop, arguing that this was engineering work, that he simply needed to verify that his tape recorder was working. They let him continue behind a movable partition, guarded by a security officer. But the press did find out, and they began to push into the room: the pencil-and-paper reporters, the television broadcasters, the radio men. With the gaps in the strips, the 200-by-200-pixel square frame was elongated into a rectangle, but soon, the edge of the planet was clear. *Mariner 4* had taken the first close-up picture of Mars.

Even though the image was half planet, half blackness of space, it was still hailed around the world. The largest-circulation French daily printed the first of the final images, once it was rendered by the computer, across five columns of its front page. IT'S MARVELOUS, read the huge banner of *The Evening News* of London. As soon as the picture was placed into the hands of Pope Paul VI, he wrote across it, *vidimus et admirati sumus*—"we saw and we gazed in wonder."

The camera had fog in it and some of the scan lines failed, causing streaks across the frame. "The resolution was awful," recalls JPL engineer John Casani. "You really couldn't see much." But the images would presumably get better as *Mariner 4* came closer and closer to the planet, imaging it as the sun struck the landscape more obliquely, picking up more contrast.

Among the streaks in the first two images, one dark area appeared to be real. It was twenty kilometers wide, shaped somewhat like a W. With the arrival of the third image, other possible features were identified, including a smaller smudge, just three kilometers across. Low hills perhaps? The first three images were released to the press for a quick look. As for interpreting the images, the center director urged patience. He reminded the public that the team's collective human strength was reaching its limits.

Leighton began utilizing some electronic tricks to improve the quality, like erasing the clearly aberrant lines that arose from faulty scanning. But when he got to frame seven, he stopped in his tracks, struggling to believe what he saw. He called Jack James, the mission director, and the then project manager, Dan Schneiderman, into a small, secure room and showed them the tiny Polaroid of the video scope. It wasn't at all what they had expected. They stared at the image in quiet disappointment. Eventually, Schneiderman uttered what they all knew to be true: "Jack, you and I have a twenty-minute jump on the rest of the lab to go out and look for new jobs."

SCARCELY ANYONE HAD been prepared for what frame seven revealed, much less what they saw in the next dozen images. "My God, it's the moon," thought Norm Haynes, one of the systems engineers. There were craters in the image, all perfectly preserved, which meant the planet was in bleak stasis. The crust hadn't been swallowed by the churn of plate tectonics, but, more important, the surface hadn't been worn down by the ebb and flow of water. Preserved craters meant there had been no resurfacing, no aqueous weathering of any kind resembling that of the Earth. As with the moon, it appeared there had never been any significant quantity of liquid water on the surface—no rainfall, no oceans, no streams, no ponds.

Stunned, the *Mariner 4* team didn't publicly release the images for days as they tried to understand the implications of what they were seeing. Finally, they scheduled a press conference. Lyndon Johnson, who had been following the spacecraft closely, hosted it at the White House. Just a few months earlier, he'd made the mission a centerpiece of his inaugural speech, addressing a country still reeling from John F. Kennedy's assassination. He'd asked the crowd to think of their world as it looked from the rocket hurtling into space, how it was like "a child's globe, hanging in space, the continents stuck to its side like colored maps." He asked them to imagine their fellow passengers on a dot of Earth, to realize that we all have but a moment among our companions. Now, with the results of *Mariner 4* in hand, that dot of Earth felt more isolated than ever before.

When Leighton took the podium in the East Room on July 29, two weeks after the flyby, he explained how man's first close-up look at Mars had revealed the fact that large craters covered at least part of the surface. "A profound fact . . ." he said somberly, as his head swayed slightly to the left. He read from his notes about the nearly seventy craters in the images, ranging from five to 125 kilometers in diameter. Leaning into the microphone, he described how the density of their occurrence was "comparable to the densely cratered uplands of the moon." It was "a scientifically startling fact."

Upon seeing the pictures, Lyndon Johnson sighed, "It may be—it may just be—that life as we know it . . . is more unique than many have thought." The mission's instruments also revealed that the air on Mars was terribly thin. The pressure was minuscule—only a few thousandths of the pressure on Earth—which helped explain why the incoming meteors hadn't burned up. The tiny whiff of carbon dioxide that had been detected with a spectrograph from Earth, and had been assumed a trace constituent, turned out to be essentially the entire atmosphere on Mars. The ground temperature was a frigid minus 100 degrees Celsius, and there was no evidence of any kind of protective magnetic field. The images were pockmarked all the way until they fell off the face of the planet.

The reality of the cold, hard, desolate world was beyond anything that scientists had imagined, beyond even the imaginations of the great science-fiction writers. "Craters? Why didn't we think of craters?" Isaac Asimov, upon seeing the *Mariner 4* images, reportedly asked a friend. The possibilities for the planet had disintegrated, our wild imaginings abraded to nothing. Humanity had spent centuries envisioning Mars as similar to the Earth, but Mars was bombarded, blighted, empty. On July 30, *The New York Times* declared, dispiritedly, what those at the press conference had struggled to say for themselves: Mars was probably "a dead planet."

CHAPTER 2

The Light that Shifts

THERE'S A TINY print of one of the *Mariner 4* images on the wall of my office. I have it upside down and tilted, ever so slightly, to reflect true north. The picture is black and white, bordered with hatches. In the image, the sun shines down at an angle over the ragged Martian surface. It lights half the rim of each crater, then shadows the opposite side. I affixed it to the wall next to my desk because it speaks volumes about the challenge of doing science on other planets. The ground is visible in that grainy image, but it's blanched of color, distant and barren. I know that it's Mars, the vast terrain south of Amazonis Planitia, but at the same time, it's nothing like the Mars I know.

I have it hanging next to a copy of the 1962 Mars map used to design the *Mariner 4* mission. It was drawn up at the behest of the U.S. Air Force, and it's the same map that hung throughout the corridors of JPL a half century ago. The contrast couldn't be more striking. Mars is smooth, awash in creamy pastels of peach and gray. Tracts of bright and dark areas are graced with names that curve and slant to fit the landscape: Thaumasia arcs over Solis Lacus; Mare Hadriaticum is horseshoed around the bowl of Hellas. Above and below the rectangular Mercator projection of the planet are smaller views of the curved Martian globe, six in all, floating against the blackness of space like a collection of holiday ornaments. Whereas the *Mariner* image is a set of static pixels, lingering alone, the planning map is a representation of a world. It's hypnotic, suffused with meaning. Every position, every orientation, every shape, every shading—each indelibly captures a human interpretation of an observation.

I know what it's like to make a map. I learned to survey in the eastern Sierra as part of a winter field camp out in the no-man's-land between Death Valley and the Mojave Desert, where the roads cut like scratches into the bleached and bending landscape. It was a swath of the Earth that no one really needed to know anything about, almost by definition. But that was the point: We were there for the challenge. The goal was to reduce the million-year-old pushing and pulling across a wild expanse of Earth's crust into a set of neat lines, etched on flapping pieces of map paper.

Out in the death-dry mountains, I slept in a small yellow tent, rising with the sun each morning. I'd throw on a thick old sweatshirt, eat breakfast from a bent metal cup, grab my Brunton compass, and meet the other graduate students by the trucks. The air was clear and cold and still, making things that were far seem close. We'd drive into the distance and spend the day tramping around the desert washes and ancient riverbeds, ascending the rock slopes.

I was learning from one of the greats—a tectonicist named Clark Burchfiel, who had a gap-toothed smile and learned to play football back in the time of leather helmets. He'd recognized the pull-apart origin of Death Valley some forty years earlier. He trained us to stick closely to outcrops, places we could measure a strike and dip in the folded and faulted rock. I used my rock hammer to break off small chunks of the outcrop to inspect the minerals, swinging it over my head to get enough torque, then recoiling as it made a sound like the smashing of teeth. I marked the GPS coordinates where molten quartz had once squirted through sills. I traced where the brittle, jagged rocks ceded to pebbled pavement, then to alluvium.

One evening, Clark tossed me a rock. The minerals seemed to

have ripped, lacerated in the darkness of our planet's twisting and sheering. "This rock has seen the face of God," Clark whispered, seemingly to himself. As I stared at it, I realized I'd spent my life walking on top of the thinnest of eggshells, oblivious to the heat and the pressure beneath me, oblivious to the magnitude of forces and depths of the physical world. Yet as hard as I searched for fault lines and intrusions on my own, I passed my days in the field wandering around feeling lost. The desert was unyielding and silent. There were great boulders perched upon ledges, and sometimes, with no one around, I'd push one with all my might until it would trundle off the peak, just to hear it crash down hundreds of meters below me, to watch it break open.

I'd work from dawn until dusk, struggling to make sense of my measurements, thinking that these mountains were no place for a beginner. Then the sun would set. I would turn in as soon as the fire died, as darkness settled over the Mojave. I'd switch on my headlamp, illuminating my little yellow dome against the cold playa. I'd lugged a dozen books with me. They were stacked in piles around the inside of the threadbare tent, which held next to nothing else. I read *West with the Night* by the pilot Beryl Markham, who flew passengers around Kenya in the 1930s for a shilling a mile. Ernest Hemingway called the book "bloody wonderful" in a letter to a friend: "This girl . . . can write rings around all of us." I read Michael Ondaatje's *The English Patient,* and like the protagonist Ladislaus de Almásy, who brought Herodotus's *Histories* into the desert, I began taping maps and sketches into the pages of my books. Cather, Dostoyevsky, Dillard, Blake, Coetzee, Stendhal. I searched them like a nautical almanac. All I wanted was to find some solid points, some method to triangulate, some way to pattern a sense of human understanding onto the vast physical world around me, a world marked by human absence.

Soon, though, I began to realize the Granite Mountains weren't as intensely empty as they seemed. When I'd first gazed into the Mojave, everything seemed muted. All the color had been drained, sipped away by the parched air. The plants were a whitish khaki green, like fistfuls of dried herbs. I had the urge to spit on them, thinking it was the least I could do, a small act of kindness. But after a while, my senses started to adjust. The sagebrush began to look like splashes, almost like raindrops hitting a lake. I started to see the life all around me—in the spine-waisted ants and blister beetles, even in the dark varnish of the desert rocks, a sheen potentially linked to microscopic ecosystems.

One day, I traced where a fossil layer had vanished, the erasure of creatures that had dominated the world for a twinkling of history. Then another afternoon I noticed similar shapes several kilometers away, in similar rock. I followed how the belts of rock plunged into the earth, then scanned the horizon, trying to envision where they might reappear. I was connecting the dots through the depths below, the bending masses, hundreds of meters beneath my boots.

The effect this had on me called to mind another of the books in my tent. I'd read about Antoine de Saint-Exupéry and how his plane was forced down in the Sahara one night. Helpless until day dawned, he fell asleep on a hillock of sand, then awoke suddenly on his back, "face to face with a hatchery of stars." He was "seized with vertigo," flung forth as if he were falling, as if the sky were a sea and he was diving headlong into it. The day on the ridge felt like that, the convex suddenly concave. I had a visceral sense of the world popping from two dimensions into three, of seeing a landscape in a way I'd never viewed it before.

With a little data and a little imagination, I was beginning to grasp how disparate observable strands could be woven into a system. And once the terrain began to make sense, all I wanted was to get to the next ridge, to fit it into my map. It was as if I could peer into the unseen ground. I didn't want mere pieces of that mighty system, some surficial understanding. I wanted the coherent whole.

THE MARS PLANNING map that hangs in my office is full of cartographic detail—a panoply of features and names. But if you step back a bit, what first catches the eye is the light and the shadow. All across the map are hubs and spokes: an extensive interconnected system of perfectly straight lines. The crisscrossing patterns are the color of smoke. They're not black, not unambiguous, but they are impossible to miss. They decorate the surface like a Victorian lace collar.

The lines date to the late 1800s, when they were first extensively recorded by Giovanni Schiaparelli, an astronomer in Milan who would forever change our vision of the planet. There had been much talk of Mars in the summer of 1877 as it swung particularly close to the Earth. An American astronomer had just discovered two moons through the great glass of a sixty-six-centimeter telescope at the U.S. Naval Observatory in Foggy Bottom. The lens of Schiaparelli's telescope was much smaller, only twenty-one centimeters, but it was made out of high-quality glass, so he decided to see whether his instrument might also be suitable for observing planets. He climbed to the rooftop of the Brera Palace. A terrible storm had passed through, and he was struggling with the conditions, unable to resolve double stars in the windy, cold air. At just before 10 P.M., with one eye to the telescope, the other on his notebook, he made his first ever sketch of Mars: a circle near the book's binding, a small white space denoting the polar cap, an apron of shading descending down from its edge, and, finally, a distinct round spot within a crescent of darkness. He noted in his observations that he couldn't find the feature on an existing British map of Mars, considered the most accurate in the world. Puzzling. But then again, the air was not good.

Schiaparelli observed Mars the next night, and the night after that. The more he observed the planet, the more he was confused by the British map, which looked like two cartoon hands of dark terrain rising up from the equator. The dark areas were designated as seas, and the light areas continents. But nowhere among Kepler Land, Dawes Ocean, Herschel Continent, or the De La Rue Sea were shadings that actually matched the ones Schiaparelli saw. In fact, none of the maps of Mars seemed to bear any resemblance to the planet he saw through his telescope.

Given the wobbling atmosphere, Schiaparelli had to work fast, recording the image he saw before the vision of it faded. But he utilized a tool that few of the other mappers of Mars had: a tiny micrometer. He affixed the miniature contraption, which he'd learned to wield in Russia, to the eyepiece of his telescope. This helped him to locate several dozen points of longitude and latitude; he could then use these to quickly orient himself. Schiaparelli's preternaturally sharp vision and this dedication to form resulted in a breathtakingly specific new map of the planet.

During those nights up on the rooftop observing Mars, Schiaparelli noticed curious features crisscrossing the planet's surface, connecting dark patches—lines that would entrance and bedevil scientists for decades. He interpreted each dark patch to be a sea, "the saltier the water, the darker it appears." He conjectured that the lines linking them were waterways. In time, he identified dozens of these "canali." They always originated in a dark patch and terminated in another dark patch or another one of the canali, never in the middle of a landmass. In some cases, canali even appeared to split, changing rapidly into two parallel canals, closely spaced.

Within a few years, a French astronomer named Camille Flammarion seized on Schiaparelli's maps and began interpreting them in the most optimistic light. In Italian, the word simply meant "channels." Schiaparelli, who had trained in civil architecture and hydraulic engineering, thought that the features might be straits, like the English Channel or Mozambique Channel. But the word rushed into the wider world as "canals," with everything that implied.

In *La Planète Mars et ses Conditions d'Habitabilité,* Flammarion noted how the "canals" didn't meander like streams and rivers. In fact, the canals on the maps seemed oddly geometric. Were they public works? Across the face of the Earth, technology was manifesting itself in new ways. The Erie Canal, dubbed the Eighth Wonder of the World, was completed in 1825 and twice enlarged in the second half of the nineteenth century. France had been heavily involved in the creation of the Suez Canal, a maritime shortcut around Africa that opened in 1869. In 1881, the French also commenced work in Panama on a new passage to link the Atlantic with the Pacific. Flammarion analyzed the canals in the context of other observations and concluded in a massive compendium of all the sketches of the planet made through

1892 that "the habitation of Mars by a race superior to ours seems . . . very probable." What else could explain the scale and regularity of the Martian canals?

It was the heir of a textile fortune who brought the canals to America and who would become their loudest and most prominent proponent. Percival Lowell had grown up in a grand mansion on Heath Street in Brookline, a Boston suburb. The house was nicknamed "Sevenels," for the seven Lowells living there. Lowell's brother would become a president of Harvard and one of his sisters a famous poet. Like his siblings, Lowell had access to a vast fortune. Upon graduating from Harvard, where he studied mathematics and dabbled in astronomy, he left on a customary grand tour and later spent several years traipsing through the Far East on various cultural and diplomatic missions. He was a self-proclaimed "man of moods," one moment jocular, holding court among friends in his piazza, the next somber, alone, chain-smoking cigars. He liked tennis and walking, but not golf, not motoring. He owned one of the fastest polo ponies in America. He had tremendous personal magnetism and typically struck others as boyish and eager, though he was a hermit at heart.

Upon Lowell's return in 1893, his aunt Mary presented him with a copy of a *La Planète Mars et ses Conditions d'Habitabilité* as a Christmas gift. Fluent in French, Lowell devoured it, electrified by the Schiaparellian network that Flammarion had interpreted as a series of waterways flanked with vegetation, carved deep into the surface to irrigate the land. The scope of the excavations was simply dazzling. There was tangible evidence of a civilization at hand. What could this be if not the biggest discovery ever made, if not a natural extension of the Copernican Revolution? Inside the book, Lowell scrawled, "Hurry!" The family motto was *occasionem cognosce*—"recognize opportunity"—something Lowell had become quite good at. He knew that in just a few months Mars would swing into opposition—when Mars and the sun aligned on opposite sides of the Earth. It would be Mars's closest approach to our planet in fifteen years, and he wasn't going to miss it.

In January, he met with a rugged young astronomer named William Pickering. Pickering had observed the linear features on Mars himself, reporting on them in an 1890 article in *The Sidereal Messenger*.

He had also just returned from a remote outpost of the Harvard College Observatory in Peru, where he had developed his "Standard Scale" for rating astronomical vantage points. Pickering convinced Lowell that optimal viewing conditions could be found in the Arizona Territory. Smog and light pollution, by-products of industrialization and urbanization, were becoming major problems, so Lowell quickly set out to build an observatory "far from the smoke of men." He decided on a high mesa at Flagstaff, a place where the atmosphere was steady and the night was cloaked in deep darkness. As Lowell insisted, "the best procurable air."

Pickering designed a prefabricated dome that was shipped west by rail. Ground was broken that April, and the first observations of Mars were made in May. Not lacking for funds, Lowell's Arizona outpost soon acquired a beautifully crafted sixty-centimeter refractor from Alvan Clark and Sons in Cambridge, Massachusetts, the leading optical manufacturer of the day. Alongside the telescope Lowell affixed a ladder assembly, atop which he set a kitchen chair. From his solitary perch, "but one watcher, alone on a hilltop with the dawn," he scrutinized the planet through his state-of-the-art lens, sketching elaborate maps of the Martian canals, rendering them as perfectly distinct lines. Confident in his exceptional optics and eyesight, from an observation post reputed to be as good as our planet could provide, he identified dozens more canals than Giovanni Schiaparelli ever had.

Lowell patiently mapped the entire planet, including even the dark patches of the surface that had long been understood to be oceans. The canals were everywhere, even there. Lowell conjectured that he was not looking at a planet like Earth, awash in salt water, but rather at a world that had lost its seas, a world where the rain had stopped. Thus the need for a global network capable of pumping what precious water there was from the melting polar snows toward the equator each spring, giving rise to dark patches of vegetation. "If . . . the planet possesses inhabitants, there is but one course open to them in order to support life. Irrigation, and upon as vast a scale as possible, must be the all-engrossing Martian pursuit." And since the canals were Mars-wide, and since there were no visible boundarics between regions or nations, Lowell reasoned that the planet had probably reached a kind of geopolitical end state where a group of benevolent oligarchs had come to direct the social order.

Whatever the likelihood of his political speculation, Lowell's telescopic evidence struck the world as wildly exciting news. He didn't have an advanced degree, but neither did many prominent astronomers at the time. There wasn't all that much to learn in graduate school, because there wasn't all that much to teach. Lowell courted the scientific journals and newspapers, and copies of his book *Mars* flew from the shelves. With its publication, much of the literate world was persuaded that proof of intelligent life had been discovered on the Red Planet. In English, French, and German, he addressed large crowds in North America, Paris, and Berlin. A "brave and brilliant *début* for the new science," heralded the *Boston Evening Transcript*. The public's fervor exploded as professionals and amateurs alike crowded in to hear Lowell's talks and raced to purchase telescopes.

But Lowell had just begun. He soon started to construct a narrative for Mars's history, a chronicle of events to account for how Mars came to be. It fit perfectly with his interpretation of planetary formation and the idea that planets would march toward an evolutionarily advanced state—both physically and biologically.

As an undergraduate at Harvard, Lowell had completed a thesis on the nebular hypothesis. The theory, first suggested on a somewhat intuitive basis by the philosopher Immanuel Kant during a foray into astronomy, and later by the celebrated French mathematical astronomer Pierre-Simon Laplace, held that rings of gas shed by the cooling, contracting sun condensed to form planets. Since entropy—the tendency toward disorder—was unidirectional, it would eventually lead to the senescence of the solar system, with the smaller planets dying first. So, from the time of their birth as molten masses, Lowell reasoned, the planets progressed through stages of development. Mars was clearly in a terrestrial stage where oceans had disappeared, but it was rapidly approaching a dead stage, an airless stage, the stage of Mercury and the stage "so sadly typified by our moon, a body now practically past possibility of change."

The implications for our own planet were not lost on Lowell. Mars had advanced to a state that the Earth too would reach. Earth was still in a terraqueous phase—its sedimentary rocks laid down by water—but its fate was sealed: "The outcome is doubtless yet far off, but it is as fatalistically sure as that tomorrow's Sun will rise, unless some other catastrophe anticipate the end," Lowell wrote. "It is perhaps not pleasing to learn the manner of our death. But science is concerned with only the fact, and we have Mars to thank for its presentment." Mars, he realized, was giving us a glimpse of our own future.

Lowell's tour de force of popular science would hold sway over the public for years. Yet slowly, quietly, hints of doubt about the canal theory began to emerge, mostly in foreign periodicals. As early as 1894, a British solar astronomer had noticed that series of tiny sunspots tended to be drawn by the eye into lines and wondered whether the same thing might not be the case with small details on Mars. In 1903, he arranged a simple demonstration to challenge Lowell. He asked a group of schoolboys at the Royal Hospital School in Greenwich to copy a series of black-and-white drawings, placed at the front of the classroom, on which dark dots had been inscribed. The boys at the desks near the front mostly drew dots, but the boys at the back of the room drew lines: For them, the dots appeared to merge together. Lowell, never one to be dissuaded, countered that linear features would also appear as lines at great distances.

Taken aback by the challenge to what he viewed as his most important scientific legacy, Lowell turned to photography, hoping it would quiet the controversy. Because of its ability to build up an image of a faint object over time, photography was a tremendous asset in stellar astronomy. It had also revealed new moons around the outer planets, but pinpointing the dot of a distant object and resolving its features were very different challenges. Even under perfect atmospheric conditions, painfully slow film speeds meant that details in the images of Mars would show up blurred. Nevertheless, one of Lowell's assistants designed a new planetary camera and took on the challenge. The photographs that emerged from the fixing solution in the Lowell Observatory's darkroom after the 1905 opposition were half a centimeter across, hardly the kind of images that allowed for a comprehensive examination of minute terrestrial features, but Lowell still distributed them widely. He heralded them as foolproof evidence that the canals were real, and the president of the British Astronomical Association followed suit in 1906, proclaiming that the photographs proved the "objective reality of the canals."

With much fanfare, Lowell announced that he would finance an expedition to the Andes to collect even better photographs of the planet in 1907. The much-hyped expedition was led by the well-known Amherst astronomer David Peck Todd, accompanied by his wife and Earl Slipher, a recent astronomy graduate from Indiana University. Amherst's forty-five-centimeter refractor, weighing seven tons, was shipped from New York to Chile via the just-opened Panama Canal. It was set up in the open air in the nitrate-mining town of Alianza, some seventy kilometers inland from the old port city of Iquique. While Todd made visual observations of the usual kind, showing canals, the true master proved to be young Slipher. Even though he had only trained in Flagstaff for a few months, he quickly figured out the planetary camera and over the course of six weeks took nearly seven thousand photographs of Mars. The gelatin-emulsion glass plates, laced with silver salts, were crated up and returned to Flagstaff.

Reproductions of the images, grainy as they were, were published a few months later in *The Century Magazine*. Anticipating that readers were unlikely to be impressed, Lowell insisted on including the disclaimer that the photographs were three steps removed from their original negatives, having undergone photographic printing, halftoning, and press printing. Despite his reassurance to readers that on the original negatives "[the canals] are there, and the film refuses to report them other than they are," they could hardly be seen.

Things continued to unravel. The same year, the celebrated British naturalist Alfred Russel Wallace, who independently conceived of the theory of natural selection, launched an attack on the concept based on his own research, arguing that Mars was likely too cold for liquid water and that a planetwide irrigation system was an absurdity. In 1909, the Greco-French astronomer Eugène Antoniadi, a long-time supporter of Lowell, published a map of Mars without any canals, practically the first such depiction in twenty-five years. Based on his own observations using the largest refractor in Europe, Antoniadi had changed his mind, concluding that only the "natural agencies of vegetation, water, cloud, and inevitable differences of colour in a desert region" were needed to account for the various phenomena on Mars. And all the while, the pioneering psychologists Sigmund Freud and Carl Jung were crisscrossing Europe and the United States, lecturing about the role of the unconscious, likely raising suspicions that observers might be seeing a vast network of canals on Mars because of an underlying desire that the canals existed. As Lowell's evidence for an advanced society on Mars withered, his own discipline began to shift beneath his feet. Einstein's theory of special relativity had been published, and space science swerved toward astrophysics, slowly relegating planetary science to a backwater it would not emerge from in Lowell's lifetime.

Lowell continued to write and lecture, seeking to inspire students as he became more and more marginalized from the scientific mainstream. He died of a stroke in 1916. In a moving tribute, his secretary described him as "filled by the warmth of his fire; thrilled by his achievements, with eye single towards the discovery of 'the light that shifts, the glare that drifts'—which is truth itself."

AND YET THOSE lines that Schiaparelli had documented and that had so consumed Lowell continued to haunt Mars science. Every time the planet swung into one of its biennial oppositions, Lowell's assistant Earl Slipher took photographs of Mars, one after another after another, from his early twenties until his eighties. Photographs from Flagstaff, Chile, South Africa. On countless nights, he stood and sat at the eyepiece of large telescopes, sometimes huddled in a plaid flannel overcoat, switching the plates, clicking the shutter, switching the plates, clicking the shutter, and so on as the hours passed.

Over the course of his lifetime, Slipher took over one hundred thousand images of Mars. In 1962 he arranged the best of them, sometimes alongside sketches he had made, into a volume he entitled *The Photographic Story of Mars.* "A vast collection of facsimiles and information has been amassed," he wrote in the foreword. It was the compendium of his life's work, and it became the basis of the planning map that hangs in my office, put together by the U.S. Air Force that same year.

In the early 1960s, of course, hardly anyone still believed that the lines on Mars were the handiwork of an intelligent society, but no one could say for sure what they were. Nearly seven decades after Lowell stumbled across Flammarion's treatise, the former head of the Royal Astronomical Society of Canada wrote: "The so-called 'canals' of Mars . . . have been plotted in various forms by most observers of the planet . . . I can sum up the present situation by saying that there is general agreement on the reality of the canals, in other words that they are not illusions, but result from something on the Martian surface that produces the effects drawn by the visual observers and recorded on the photographs . . ." Samuel Glasstone affirmed, "There definitely appear to be a number of linear features on the surface of Mars," in *The Book of Mars,* a NASA special report in 1968.

What were those eerily straight lines? Fissures, perhaps, caused by the drying and cracking of the Martian surface? Depressions? Radial streaks of material ejected from craters? It was difficult to explain why *Mariner 4* didn't see the spidery black webs, if indeed they were there. Did the flyby just miss them, with its handful of images only covering 1 percent of the planet? The improvement in image quality for *Mariner 4* over telescopic photographs was so striking that the mission scientists had struggled to correlate them with any of the existing pictures of Mars. Perhaps the images were simply taken at too close a range? The mystery of the linear features was one of the reasons NASA decided to launch a second pair of flyby missions to Mars, in 1969. "Photography from the mission is expected to settle this point," explained the NASA press team. In order to improve the coverage, far-encounter images were added to those taken up close during the flyby. A probe called *Mariner 5* had been sent to Venus, so NASA named its next Mars missions *Mariner 6* and *7*.

By February of 1969, *Mariner 6* was tucked inside a glistening nose cone at Cape Canaveral. On Valentine's Day, the *Mariner 6* ground crewmen began a routine test procedure. Unlike *Mariner 4*'s slimmed-down, simple rocket, the Atlas-Centaur rocket was enormous, towering over the cape like a skyscraper. Ten stories tall, it weighed 150 tons. It was far larger than NASA needed to get to Mars, but it was widely available, having been built in bulk to ferry NASA's much heavier spacecraft to the moon.

Suddenly, the sound of buckling metal echoed across the launchpad, followed by the piercing wail of an evacuation siren. When the ground crewmen looked up, they couldn't believe their eyes: The fixed solid launch vehicle, the unshakable, unyielding locomotive of a rocket that was designed to survive a blast into outer space, was collapsing under its own weight, its smooth metal skin rippling and folding like fabric, right in the middle of the launchpad.

Before they could respond, the top of the rocket had tilted to a twenty-degree angle, forming a kink in the cylinder. The balloon tank of the rocket relied on pressure to maintain its rigidity, but the design of the Atlas-Centaur had done away with all the internal structure, to save weight. The crewmen realized that the main valves must have somehow popped open, letting air rush out from fifteen-centimeter openings.

Time seemed to stop as the wires strung to the far side of the launch vehicle pulled taut, then popped. One of the workers rushed to secure the locking bolts as the booster sagged perilously toward the umbilical tower, then thudded onto the platform. Another worker wriggled his way up into the thrust section, trying desperately and finally succeeding in closing the valves, which prevented the crumpled rocket from crashing to the ground.

Those two members of the ground crew received Exceptional Bravery Medals from NASA for saving the *Mariner 6* spacecraft, which was carefully extracted from the nose cone and transferred to another Atlas-Centaur. I wonder what it must have been like watching those straight edges warp, looking up at a rocket that only seconds earlier had been firm and solid, trying desperately to orient myself.

How can a mountain lean, how can a palisade collapse, how can an Atlas-Centaur ripple? And yet *Mariner 6* still launched on schedule, followed a month later by *Mariner 7*.

TWO DAYS BEFORE reaching Mars, the telephoto shutters on *Mariner 6* opened, and the excitement began. As the spacecraft hurtled in from a hundred and fifty kilometers away, a long-studied linear feature was immediately spotted: the "canal" Coprates. It appeared on the terminator—the line that separates the illuminated day from the dark night—then arced across the disk of the planet and disappeared behind the edge. But by the time *Mariner 7* pulled within twenty-four hours of its near encounter, shutters whirring, it seemed evident that Coprates was just a collection of dark dots aligned somewhat east to west, and not even straight.

It was the same for the images that followed, as *Mariner 6* raced past the planet, and *Mariner 7* followed with a ribbon of images over the South Pole. There was no geometric pattern on the surface. No doublings, no diagonals. Not even the soft angles of a crocheted blanket. It took only eight days for *Mariner 6* and *7* to kill the linear features. The schoolboys had been right. For all that time, the lines we'd seen simply weren't there.

So it was with the whole mission. As the probes collected images with their near- and wide-angle cameras, nothing was what it at first appeared. Mottled areas resolved into crater fields. A "lump" on the southeastern limb of the planet turned out to be a detached haze layer. The W-shaped clouds that had been observed for years were not clouds at all but rather a real surface feature.

Other boundaries once thought to be smooth and regular—along the edge of the south polar cap, between Mare Cimmerium and Aeolis—proved in fact to be ragged and broken. There were uncratered expanses, like in the depths of Hellas, a smooth, featureless terrain that must have been resurfaced somehow. There were fields of chaos—jumbled, blocky, broken terrain with no known counterpart on the Earth or the moon. There was structure to the atmosphere, layers and layers that hadn't been detected before. There was a dramatic south polar cap, built with carbon dioxide ice. The mission even measured some warm temperatures along the equatorial latitudes, as warm as crisp fall days on Earth. Mars was not the Earth, but it was not the moon either. It was another world altogether. The whole mission, from the crumpling of the rocket to the data the spacecraft returned, seemed to underscore one of the most fundamental things about scientific discovery: that the truth can be a chimeric thing, that the collapse of an abiding belief is always just one flight, one finding, one image, away.

TOGETHER, THE *MARINER 6* and *7* near-encounter images covered 20 percent of the Martian surface. The probes discovered scores of new features, but the images had also undone much of what we thought we knew. We now lacked any semblance of a legitimate map. As the mission team struggled to fit all the new observations together, it became evident that NASA needed another Mars mission, a true mapping mission. And in 1971, Earth and Mars would align when Mars also happened to be at its closest approach to the sun. It would be a particularly favorable time—with Mars close in, there would be less distance for the spacecraft to travel. So in the run-up to the next launch window, NASA readied a pair of twin mappers: *Mariner 8* and *Mariner 9,* identical spacecraft that would go into orbit instead of just flying by the planet. They would establish the Martian geoid, the reference grid from which all points could be identified: latitude, longitude, altitude. *Mariner 8* would map the planet's fixed features—the permanent ones, like deserts and craters—while *Mariner 9* would map variable features, the ones that shifted with the seasons. It would do this by entering into a different orbit to image the same places at the same time of day as the Martian year progressed. Between the two missions, NASA hoped to eclipse all the historically crude and static representations of the planet's surface.

As *Mariner 8* and *9* were being fueled, enshrouded, and mated to their rockets, the Soviets readied three Mars missions of their own at the Baikonur Cosmodrome, vying with the Americans to place the first Martian satellite. Despite eight attempted missions, they had yet to achieve any success on Mars.

Mariner 8, the first of the five, blasted off on May 8, 1971, the very moment the launch window opened. Everything appeared to go smoothly for the first few minutes, but then the upper stage of the rocket began to oscillate and tumble. The payload was prematurely cut loose, and *Mariner 8* fell through the dark sky. It splashed into the sea and continued falling, all the way down to the bottom of the ocean.

The next day, the Soviet mission *Kosmos 419* launched, but the upper stage of its rocket also had trouble, failing to fire the second time, marooning the spacecraft in low Earth orbit for two days before reentering the Earth's atmosphere. It turned out that an eight-digit code to fire the engine had accidentally been issued in reverse by a human operator; embarrassing, but easy to fix. Over the next couple of weeks, the Soviet missions *Mars 2* and *Mars 3* soared into space on Proton rockets.

Mariner 8's failure had also been caused by something small, an integrated-circuit chip, no bigger than a sunflower seed. A faulty diode had likely failed to protect it from a voltage surge. The *Mariner 9* engineers persevered through the rest of May after happening upon a second issue: a short in the propellant system. They hurriedly pulled the spacecraft down from the Centaur, fixed it, then reran a full set of tests. Finally, NASA got the spacecraft back on top of its rocket. On May 30, on a clear blue evening, *Mariner 9* lifted into the sky, joining the race to make the first full map of Mars.

CHAPTER 3

Red Smoke

SCIENTISTS AT THE Republic Observatory in South Africa were among the first to notice the haze that had begun to spread across Mars. Throughout the summer of 1971, as *Mariner 9* journeyed to Mars with the two Soviet probes, those scientists had been peering up from the dome on St. Georges Road, carefully scrutinizing the planet that the three spacecraft were approaching. On September 22, they saw a bright-yellow streak starting to form. It was along the edge of Noachis Terra, a giant southern landmass, part of the heavily cratered highlands. They tracked it as it elongated, first as a thin line, then fattening into a continuous belt of clouds: the beginning of a dust storm.

Within five days, the storm had spread from east of the Hellas basin clear to the south of Syrtis on the other side of the planet. It grew, then retreated, and then suddenly, just weeks before *Mariner 9*'s arrival, the dust engulfed the whole surface. The features of Mars vanished almost entirely from view, as if the planet were wrapped in a smooth, lacquered cloud. "It looked like a billiard ball," recalled

Norm Haynes, a member of the *Mariner 9* engineering team. "We couldn't see a thing."

Slowly, a kind of panic came over the team. It was an astounding tactical problem for a spacecraft designed to study the terrestrial features of Mars. The mission was only meant to last three months after it entered orbit. In the early weeks of November, as the spacecraft drew nearer and nearer to its destination, the surface remained completely obscured. Six days before reaching the planet, *Mariner 9*'s television cameras switched into calibration mode and pointed at Mars. The images that came back were still nearly blank. The mission team reprogrammed the computer system to conserve its data storage. *Mariner 9* would circle Mars, waiting, hoping that the skies would clear and the planet would gradually come back into focus.

The Soviets, however, didn't have the same luxury, as their software was not reprogrammable. Their two orbiters, arriving just two weeks after *Mariner 9,* both snapped their pictures immediately, returning images of nothing more than impenetrable dust clouds. Like matryoshka dolls, the Soviet orbiters carried small landers but were unable to delay their release, and the landers were promptly sucked into the tempest. One did manage to land on the surface, but the only data returned to Earth were a few lines of a single, incomprehensible image. The transmission ceased less than two minutes after touchdown, before the lander could release the small tethered robot it carried, which was designed to traverse the Martian sands on a pair of skis.

The storm continued to rage. It probably started with a single delicate arc of dust, lifting off the ground like a charmed snake. Because Mars was so close to the sun, it was the peak of summer in its southern hemisphere, with solar heating at its maximum. As sunlight warmed the surface, it also warmed the adjoining layer of air. Warm air rises, and although the Martian atmosphere was thin—less than 1 percent the thickness of ours here on Earth—the rising air nevertheless drew the ultrafine dust along with it as it lifted into the sky. And as more and more dust filled the air, it began to act like a cloud of tiny mirrors, reflecting and scattering the sunlight. As the sunlight bounced, the surface cooled, but the atmosphere warmed, driving breathtakingly fast winds, churning up even more dust from the surface, and creating one of the longest-lasting, most violent dust storms that has ever been observed in our solar system, even to this day.

I wasn't alive when *Mariner 9* reached Mars, but whenever I look at those images of dust shrouding the entire planet, I can almost feel the particles choking my lungs. The summer after my sophomore year of college, I spent ten weeks caked in simulated Martian dust. I was interning in the Planetary Aeolian Lab at NASA's Ames Research Center. When I walked in to Building N-242 on my very first day, I was struck by the titanic dimensions of the lab. It was one of the largest vacuum chambers in the world—4,000 cubic meters, larger than an Olympic swimming pool—originally built to investigate the buffeting of rockets as they ascended into the atmosphere. The space was enclosed by five walls of solid concrete, comprising a pentagonal tower. I stared up, and as the ceiling ten stories above my head slowly came into focus, I tasted blood in my mouth. When I reached for my gums, the guide I was with laughed, explaining that there was Martian dust simulant all over the ground, coating the walls like brick flour. It wasn't blood I was tasting, just the sick-sweet tinge of iron hanging in the air.

Everywhere I went that summer, I carried the dust with me. It clung to my skin, my eyelashes, my teeth. There were delicate orange stripes on the undersides of my fingernails, and even though I wore a cleanroom bunny suit, the dust would still puff out of my clothes at night. I'd occasionally spot traces of it in the crevices of the floorboards of the old house where I was staying on Stanford's campus or in the seats of the van that ferried me and the other interns over to NASA's Astrobiology Academy each morning.

The dust simulant was called JSC Mars-1A. Two years earlier, nearly ten thousand kilograms of the weathered volcanic ash had been dug out of the side of Pu'u Nene, a cinder cone in the saddle between Mauna Loa and Mauna Kea on the island of Hawaii. It was the closest thing to Mars dust that existed on Earth. We had heaps of it, sieved into various particle sizes, but it was only the finest of fine dust we would use for our experiments. This was because physical forces had thoroughly worked over the surface particles on Mars, ever so slowly fragmenting them, pulverizing the grains until they were as fine as talcum powder. The grains were cracked by cycles of freezing and thawing and rusted over by tiny chemical reactions. But mainly they were whittled by the wind. Most of those gusts were as gentle as a feather duster, but they were incessant, for billions of years.

The Mars Surface Wind Tunnel cut across the dusty floor of the giant chamber, and it was there that I set up my flow-field experiments. My goal was to examine how Martian dust was entrained in the wind and how it would settle over a spacecraft. A new mission was careening toward Mars, and in six months' time it was going to land in the layered terrain near the Martian South Pole. My project was designed to help gauge how much wind-carried dust might collect on the lander's broad, flat solar panels—to figure out how much light the dust might obscure, how much power would be suppressed.

The Mars Surface Wind Tunnel reminded me of the forts I'd made as a child—cardboard boxes lashed together with reams of tape. It was just big enough for me to climb inside. In my bunny suit—a papery-thin coverall that left only my face and hands exposed—I would crawl to one end of the tunnel, position the solar panels, then crawl to the other to sprinkle layers of dust where they could be lifted by a giant laminar-flow fan.

When everything was ready, I'd retreat to the control room with the wind-tunnel technician and the other student who was working there that summer. From a tiny reinforced window, I watched as we ran the experiments. When we flipped a switch, the steam plant across the street pulled the vacuum. The chamber would creak as the pressure began to drop. We began at one thousand millibars, normal atmospheric pressure, then lowered it to five hundred, to two hundred, to one hundred.

After reaching six millibars, I would wait a few minutes, then check the controls and begin my measurements. With the press of a button, on the tiniest gust of wind, the dust would balloon into billows. It lofted so easily, like nothing, on the smallest puff from a pressurized air jet. It would cling to the solar panels and I would dutifully record the drop in power readings. But my eyes kept wandering back to that paisley of swirling eddies, cut by shafts of light from the flood lamps above. The dust was exquisite. It filled the meager air with particles that seemed like they would float forever.

AFTER THE DEVASTATING loss of *Mariner 8, Mariner 9* would have to do the work of two orbiters. Though NASA headquarters insisted on prioritizing the mapping of fixed features, the mission team was able to work out a compromise orbit that would allow it to complete at least part of *Mariner 9*'s original mission: studying the features that were in flux, including something called the "wave of darkening," a phenomenon that had captivated Mars scientists since the nineteenth century.

The "wave of darkening," a term coined by Lowell, referred to how the terrain seemed to darken at the poles every Martian spring and progress slowly toward the equator, an event that had been observed repeatedly with ground-based telescopes. What could explain it? Many astronomers had interpreted it as a sheen of vegetation, despite the parched conditions. The darkening was peculiar in that it proceeded in the opposite direction from that on Earth. Here, vegetation grows from the equatorial latitudes, where it is warmest, toward the poles. But on Mars, where water was scarce, it was hypothesized that water would be the limit to growth. Water would become available first near the poles at the end of the local winter, as ice began to vaporize—then liquefy—spreading slowly toward the equator.

The mystery had tempted the imagination of Mars scientists for years. In 1956, a University of Chicago scientist named Gerard Kuiper noted what he thought to be "a touch of moss green" in the equatorial regions. A researcher at Harvard University performed follow-up spectroscopic studies in the late 1950s, detecting specific absorptions among different wavelengths of light over the dark areas of Mars, which were widely interpreted as organics. "This evidence," he explained in *The Astrophysical Journal,* "and the well-known seasonal changes of the dark areas make it extremely probable that vegetation is present in some form." By 1962, his French colleague was even able to establish a rate for the wave of darkening: roughly thirty kilometers a day, according to the photometers at an observatory in the Pyrenees. The bright features on Mars were deserts, to be sure, but it had been impossible to tie the dark areas to underlying geologic structures. Part of *Mariner 9*'s mission was to determine if those dark areas were evidence of life.

THE INTEREST IN the wave of darkening reflected a subtle shift that had played out in Mars science in the early twentieth century: from Mars as home to a civilized world to Mars as home to a vegetated world. William Pickering, the same man who had led Lowell west to Arizona, had played a key role in developing the idea of a botanical Mars. Pickering was an astronomer and a naturalist. He was an intrepid hiker. At twenty years old, he'd climbed Half Dome in Yosemite, and by twenty-four, he'd created the first recreational guide to the White Mountains in New Hampshire. He included descriptions of how to reach a summit in the absence of a trail and advice about the use of tar soap and pennyroyal for mosquitoes, milk for sunburn, and washleather for blisters. He spent hours gazing at dramatic escarpments. "It is then, and only then," he once wrote, "when high above the tree-line, with one or perhaps two companions, that the grandeur and loneliness of the great peaks really break on [a person]; then for the first time does he begin to understand . . ."

Pickering loved gazing into the distance, and he loved gazing at Mars. He preferred wild places for astronomical observing, where the viewing conditions were exceptional, he maintained. There was simply less water vapor to wobble the air, and fewer clouds and storms at altitude on mountain peaks. With the help of his older brother, the head of the Harvard Observatory, he led several efforts to establish remote astronomical observatories, including Lowell's in Arizona. For a time, he used this belief in the superiority of remote locations to defend the existence of the lines that Lowell believed were canals. Like Thomas Henry Huxley, who protected the theory of evolution so vigorously that he came to be known as "Darwin's Bulldog," Pickering vehemently argued that the astronomers in northern Europe and the eastern United States who had not experienced a place like Flagstaff had no right to opine about the reality of surface markings on other planets, as their observatories simply didn't allow for "good seeing." They might as well express their views on electrodynamics or physiology, he maintained, or other areas they knew nothing about.

Pickering, however, never fully bought into Lowell's explanation. He didn't deny the possibility that an intelligent civilization might exist on Mars, but he did harbor doubts that the canals were foolproof evidence of one. His observations from his outpost in Peru had nursed his hesitation, and he came up with many alternative theories.

For instance, in 1905, while he was exploring the volcanoes of Hawaii as possible analogs for the moon, he happened to notice a series of cracks in the desert extending to the south of Kilauea. Based on those volcanic cracks, he began to suspect that the Martian canals weren't waterways crossing dark patches of vegetation; rather, the "canals" were *themselves* vegetation. Waterways should have glinted in the sunlight, after all. But if steam was emerging from naturally formed cracks, that could in turn nourish plant life like trees, low bushes, and ferns. He knew that this idea wasn't entirely satisfactory either, as it postulated lots of volcanic activity on the surface of what everyone knew to be a small, cold planet. But he kept playing with the data, turning the evidence over, seeking out clearer observations and better theories, never convinced that the lines on Mars were quite what they appeared to be.

Then, two years later, in the Azores, he found himself gazing at a hill from a distance. Once densely wooded, it had been deforested in geometric patterns to make pastures for cattle. Suddenly, another theory for the linear features came to him: Perhaps the lines were a vestige of a plant species in decline? "Imagine that the whole surface of the planet was originally covered with some form of bush or tree, which in the northern and equatorial regions has now been largely destroyed," he wrote. "Its continued presence in the southern regions would account for the so-called seas, while narrow, more or less continuous, strips of it would account for the canals."

In 1911, Pickering set sail for the British West Indies, to a high plateau in the central mountains of Jamaica, with a pocketful of funds from the Harvard Observatory, determined to establish a small viewing station there. He leased a one-story plantation house that would be his base. There were no lights or running water or telephone. On a large patio once used to dry coffee beans, Pickering set up his twenty-eight-centimeter Clark refractor. In order to appease his older brother, who was working hard on a stellar map of the sky, he'd occasionally observe a double star, even though he "didn't care a whoop about the stars." It was known that many of them had burned out long ago, even though their light still traveled toward the Earth, and he couldn't be bothered with creating a catalog of useless suns. "The enormous size of our stellar system is of no consequence to us," he once wrote. "If it only contained ten thousand stars instead of a thousand million we might perhaps be just as contented."

It was beneath the telescope on that patio that Pickering developed his final theory to account for the canals: Again, it was vegetation-based. Mars's atmospheric circulation, which he conjectured to follow regular patterns, resulted in certain parts of the ground being moistened again and again by thunderstorms. The straight, narrow bands beneath those regular squalls had then become marshes teeming with life in an otherwise-silent landscape. "Had it not been settled by the Europeans," he wrote, "the United States would still be a wilderness. How much less should we hasten to accord civilization to a planet of which we know little . . ." Instead of being irrigation ditches or pseudo-canals, he concluded, the patterns, however straight they appeared, were accidental.

By the time he had devised this theory, however, it was 1914, and terrestrial catastrophe loomed in the form of the impending World War. From his secluded perch in Jamaica, Pickering seemed to take refuge in a nearly obsessive series of Mars observations, each transmitted in a monthly report. To him, Mars was a pristine wilderness. Without civilization, there could be no war, no conflict. He was pushing sixty, too old to be called up, but still with a youthful toughness about him, his boyish eyes staring out from a face creased by sun and wind. His beard, even when trimmed, had an unruly spray of hair at the chin. He refused to own an automobile, so when the time came to cable his reports, he trotted into the nearest town on one of his two horses, Jupiter and Saturn. He fancied himself less a scientist than an emissary, particularly to astronomers in the northeastern United States, a class of people "not fortunate enough to reside in those portions of the world where the seeing is habitually good." His reports were full of maps, data tables, and scatterplots, but, eager to defy convention, he didn't send them to scientific journals. He instead sent them to an American magazine called *Popular Astronomy*.

These bulletins represented the first time that the world received regular news reports about another planet. He kept it up for years, spending nights beside the plantation house-turned-observatory, recording his impressions of Mars into leather-bound books in skittish, swooping cursive. He made hundreds of pencil sketches, capturing the changes in Eden, Elysium, Arcadia, Chryse, and Utopia. They were joined by dozens of paintings, delicately colorcast in carmine and shades of sienna. All the while, he knew that the astronomical community was moving toward photographic research, using techniques Pickering himself had helped to pioneer. Yet he was undeterred, becoming increasingly convinced that the "human eye must reign supreme." It was not long before he abandoned astrophotography almost entirely in favor of basic visual observations of planetary surfaces.

In his reports, he tried his best to stick to what he saw. He laid out a taxonomy of terms. He argued that certain questions were best left "in abeyance," like the existence of Martians. Nevertheless, his vision of the landscape permeated everything he wrote. He talked of growing and receding storms. He reported on the coastlines of blue-tinted bays, the greening of the southern *maria,* and the wild, heavy rain falling in dark, uniform sheets. Sometimes, gargantuan floods surged through the north like the spring torrents of Siberia. At other times, belts of high-rising cumulus clouds swept the sky, not unlike those he'd seen above western Bolivia. He reported when the south polar cap seemed stippled with hoarfrost and when icefields appeared in the torrid zone. He announced when Mars was "snowed under" as far south as southern Labrador on Earth. In so doing, Pickering brought the world a new vision of Mars, an alien marvel that was no less glorious for its lack of a civilization. It was an expansive, unexplored landscape, every bit Earth's equal.

PICKERING'S VISION OF a vegetated Mars resurfaced before the launch of *Mariner 9,* though not exactly as his missives portrayed it. No longer was Mars imagined as having teeming marshes or a surface that was like a thickset stretch of the Amazon basin, where during the rainy season hundreds of kilometers were flooded to the depth of several meters. No longer was it understood to have a dense atmosphere, welling with storms and full of moisture and heat. Temperature measurements of Mars with vacuum thermocouples—circuits of wispy wires of different metals soldered end to end—suggested that conditions were cold, but not that cold: the thermocouples registered signs of warmth in Mars's dark areas, measurements that could be explained by the growth of simple vegetation like moss and lichen, not unlike the Siberian tundra.

Despite his deep affection for the "grandeur and loneliness of the great peaks," Pickering had believed that the surface of Mars was devoid of high mountains, a commonly held belief that also persisted right up until the arrival of *Mariner 9* in 1971. After all, Mars was far smaller than Earth. How could such a tiny interior generate enough heat to sustain vigorous volcanism and plate tectonics, the forces that build mountains? In fact, his final vegetation theory depended critically on the premise of a flat surface; without it, the atmospheric circulation patterns fell away, along with the squalls that watered vegetation, lush and wild.

For two months after arriving at Mars, *Mariner 9* continued to patiently circle the planet as the dust storm raged on. In early January, after weeks of waiting, observations showed that the dust finally appeared to be receding. Recognizable features were beginning to peek through the red haze. And when at last *Mariner 9* began its imaging, the surface revealed itself to be anything but level. The first thing to emerge from the dust pall was a sort of fuzzy spot. The imaging team, beside themselves with excitement, began poring over the classical maps to identify it. It seemed to align with the approximate location of Nix Olympica.

In the days that followed, three more spots slowly appeared, all in a line, which the team dubbed "North Spot," "Middle Spot," and "South Spot." A press conference was scheduled. In the midst of it, surprising everyone, the head of the imaging team blurted out that the spots must be the summits of wildly tall volcanoes. They simply had to be high areas, towering above the dust. He divulged that he thought he'd even seen collapsed craters, characteristic of cauldron-like volcanic calderas that form when a magma chamber empties. NASA had flown by Mars three times and hadn't seen a single volcano. In fact, *Mariners* 4, 6, and 7 hadn't spotted any topography at all on the planet's surface. In the parts of the planet they'd imaged, there wasn't a single significant shadow, not a single contour on the horizon. And if the spots were in fact mountainous volcanoes, they were far larger than any volcano on Earth.

The head of the imaging team was right: The three spots on the crest of Tharsis Rise were volcanoes with calderas. They would become known as the Tharsis Montes: Ascraeus Mons, Pavonis Mons, and Arsia Mons. Nearby Nix Olympica was reclassified as Olympus Mons, one of the largest mountains in the solar system. The existence of volcanoes—clearly the mark of a once-hot interior—was as thrilling as it was unexpected. It meant Mars must have had large inventories of gases bubbling out of magma, the same gases that originally filled our atmosphere and condensed to fill our oceans.

The volcanoes weren't the only enormous features on Mars. There were also gargantuan canyons, revealed slowly as the images rolled in. "We saw them coming," recalled engineer Norm Haynes. Seeing the pictures as they arrived "was like pulling back a curtain, little by little, day by day." Eventually, when the orbital swaths were laid down side by side, it became clear that the planet's side was split like the Great Rift Valley. A nearly 4,000-kilometer gash ripped through the equator, large enough to engulf the Grand Canyon again and again, stretching around a fifth of Mars's circumference. It would become known as Valles Marineris, the Valley of Mariner. Far from being a cratered, empty expanse, Mars was a place of enormous variation. It was just by chance that the earlier flybys had missed, well, everything.

With the settling of the dust, however, the idea of great tracts of vegetation on Mars evaporated. Mars had a wildly dynamic surface, but there was no evidence of hardy (if water-starved) primitive plants. It was true that only a few weeks sufficed to change the landscape completely, as the early studies had noted, but it wasn't the blossoming of spring vegetation. The *Mariner* team soon realized it was just the simple physics of seasonal warming on the Martian surface. As the planet tilted toward the sun, the sun heated the surface and the dust was lifted and shifted, exposing the bare underlying rocks. At great distance, this gave the impression of a bloom. The wave of darkening was just a systematic redistribution of the massive blanket of dust upon the surface, and just like that, the concept of vegetated Mars fell away, just as the idea of a civilized Mars had slipped through our fingers.

Alongside this disappointment, however, came something else, something stunning: unmistakable evidence of riverbeds. There were no linear features on Mars, no geometric lines on the surface, but the pictures clearly revealed branching forks and catastrophic outflow channels.

It was almost impossible to believe, and at first, few on the science team did. For years, evidence had been building for a dry Mars. Many on the team wondered if the apparent riverbeds could have been cut by lava, but the riverbeds weren't found only in volcanic terrain. They seemed to have distinctive meandering patterns and sandbars, the kind of features any hydrologist would instantly recognize.

But how could there be riverbeds? There was virtually no water vapor in the atmosphere. The low temperature and low pressure meant that any water that did accumulate on Mars would either freeze or evaporate rapidly. The surface pressure on Mars, six millibars, is just past the triple point for water, the ethereal combination of conditions where it coexists as a solid, liquid, and vapor. I remember seeing it with my own eyes the summer I ran the experiments in the Mars Surface Wind Tunnel at NASA Ames.

The first time I drew a vacuum, I purposely—just for fun, at the suggestion of a graduate student—left a small glass of water sitting on the lip inside the viewing window from the control room. As the air disappeared from the colossal chamber, the water in the glass slowly began to boil, then boil violently. It started splashing over the edge of the glass, splattering on the window. Then suddenly the absurd: A shard of ice appeared, right in the middle of the boiling water, bouncing around the glass like a phantom.

I understood what was happening. I'd read about the heat of vaporization in my textbooks, how a substance will cool as it evaporates, even as it boils into a near vacuum. But no amount of physics could remove the spell I was under. I gingerly touched the control boards. I looked at the graduate student, then at the young lab director, then back at the boiling ice. Is this what the *Mariner 9* scientists experienced—this profound disorientation?

There they were, looking at images that upended everything they knew, deep in a wilderness of the unexpected. What other physical laws would break, what other mysteries of the universe would burst forth from those pictures? The idea of rivers on Mars was shocking on two counts—it seemed to require a miracle to get enough water on the surface, and then it required another miracle to keep it there long enough to erode a riverbed. Even a knee-deep rivulet in the deepest reaches of the Martian surface, the places with the highest overriding surface pressure, might vanish between dawn and dusk at prevailing surface pressures.

But if the features weren't carved by lava or another viscous fluid, it meant that Mars must have been a dramatically different place earlier in its history. And to understand that would require much more than understanding the planet as it was now. It would require piecing together a complex and dynamic account of its journey through time. It was no longer the wave of darkening we needed to explain but the staggering mystery of features that looked for all the world like riverbeds. And it meant entertaining the possibility that Schiaparelli had been right, at least in part: Perhaps natural waterways had coursed across the surface of Mars. Perhaps this blighted, empty, cratered planet had once harbored life.

PART 2

A Line Is Breadthless Length.

—EUCLID'S ELEMENTS

CHAPTER 4

The Gates of the Wonder World

IN THE EARLY 1980s, not long after I was born, the miniseries *Cosmos* aired on PBS. My parents tuned in, along with millions of other American households, a public-broadcasting record. For thirteen Sunday evenings in a row, they held my sister and me in their arms on a harvest-gold couch printed with flowers, atop a brown plush rug. They adjusted the antennae of their picture-tube television until the image was clear. It was the first set they ever owned in color, and what a time for color.

The show starred a young astronomer named Carl Sagan, who rode his "Spaceship of the Imagination" through the psychedelic wonders of the universe. With the help of wild special effects, he spun among the stars, cruised through the "snowballs of Saturn," and turned his body into a silhouette of pink lasers. He jumped forward and back through time, crawled inside a giant human brain, and floated across treetops, often to a melody of trippy music. My sister and I would stare wide-eyed then doze off to sleep, to be carried up to our cribs as the credits rolled.

To say that *Cosmos* was a phenomenon would be an understatement. Sagan eventually reached into the homes of more than half a billion viewers across the globe. He'd already started building a public presence by writing for a large audience, but *Cosmos* was what made him truly famous. He appeared regularly on Johnny Carson, surrounded himself with celebrities, smoked lots of marijuana, and exasperated his colleagues. He was a turtlenecked, freewheeling prophet of science, happy to sign an autograph. For the first time since Percival Lowell, a single individual was the face of Mars science. For the last forty years, no name has been more closely associated with Mars, or the search for life generally, than Sagan. In fact, you'd be hard-pressed to find a modern scientist who cut a higher profile or had a stronger influence over the popular imagination.

A few years before *Cosmos* aired, with an iconoclastic confidence that would foreshadow his rise to stardom, Sagan daringly suggested there might be turtle-like creatures on the Red Planet. In 1974, he had submitted a short piece to the journal *Icarus,* in which he concluded that "large organisms . . . are not only possible on Mars; they may be favored." It was a tremendous leap, of course, but Sagan, like many visionaries, saw the possibilities in precise detail. These creatures, he speculated, might be plodding about with tough silicate shells to protect them from UV radiation. He conceded that because no visible vegetation had been detected on Mars, it was difficult to imagine the creatures' food source. Even so, he suggested that some among them, which he called "crystophages," might be tapping into the icy permafrost to fight the aridity, while others, the "petrophages," might be drinking hydrated minerals from the rocks themselves. Sagan pointed out that desolate tracts of land on Earth long thought to be barren of large animals were now understood to be habitats, home to polar bears and their kin. He reasoned that substantial size would reduce the ratio of surface area to volume, thereby enabling large creatures to conserve heat and moisture in a cold, dry climate. A graphic artist from *Time* magazine rendered a gigantic sprawling octopus to convey Sagan's vision to the public. It was all a stretch, as Sagan undoubtedly knew, but that wasn't his point. His point was that the case *against* large, loping creatures—"macrobes," he called them—simply hadn't been made, so why would we want to constrain our imaginations? Why not macrobes? Why not even silicon-based giraffes? As with so many things in Sagan's life, it was a highly unconventional argument, and an inspiring one. It was also an early flourish of the genius he would later show for communicating with nonscientists.

At the time, NASA was preparing spacecraft that would finally land on the surface of Mars. The 1976 *Viking* mission would conduct the very first life detection experiments on the Red Planet. Sagan was part of the mission's imaging team, and he worked to ensure that anything in the vicinity of the two identical landers would have its portrait taken in color, black and white, infrared, and even stereo. And he was up to his old tricks: When a reporter pointed out that a fast-moving creature would only "show up as a streak," Sagan didn't miss a beat. "But we can always look at the footprints," he replied.

Sagan had always had an extravagant, limitless imagination. He'd grown up in a small apartment in Brooklyn, and after reading Edgar Rice Burroughs's *A Princess of Mars*, with its Virginian hero mysteriously finding himself on the Red Planet, he'd rushed headlong into a nearby field, his arms resolutely outstretched, imploring Mars to ferry him there. As a ten-year-old, he'd sketched block-letter headlines from the future: SPACESHIP REACHES MOON!!! and LIFE FOUND ON VENUS, even a pair of astronauts advertising voyages on INTERSTELLAR SPACELINES.

He'd devoured pulp science-fiction magazines throughout his adolescence, and at fifteen he'd happened to notice an advertisement for Arthur C. Clarke's *Interplanetary Flight: An Introduction to Astronautics*. Unlike Clarke's fanciful short stories, *Interplanetary Flight* was a short technical volume, outlining everything that was known in 1950 about orbital dynamics and rocket design. Sagan was astounded by the possibilities Clarke laid out for sending probes to other planets and perhaps even sending them soon.

He raced off to the University of Chicago the next year, at just sixteen years old. The demands were high, and Sagan was nothing if not intense. He was soon suffering from chronic pain that would stay with him for his entire adult life. He drove the lonely highway to the Mayo Clinic by himself, embarrassed that he could no more than nibble at food without fear of choking. He was diagnosed with achalasia, literally "failure to relax." It was a condition of the esophagus that made it difficult to breathe and swallow, which his sister speculated was the result of his mother's obsessive, neurotic influence. The doctors tried to stretch his esophagus, unsuccessfully, just as they failed with surgery years later, leaving his lung cavity filled with blood.

Yet Sagan had a resilience that ran deep. Despite his self-consciousness and pain, he fearlessly approached important scientists to ask his questions. By putting himself on the line, he cultivated a series of brilliant mentors. He studied physics and wrote an undergraduate thesis on the origin of life, supervised by Nobel Prize winner Harold Urey. He worked in the summers with top scientists around the country, then decided to stay on at the University of Chicago to complete a PhD in astronomy and astrophysics, commuting in a blue-and-white Nash-Hudson station wagon over to the school's observatory in Williams Bay.

Although his primary PhD research was on the physical properties of planets, Sagan soon found himself at the epicenter of the nascent field of exobiology—back doing research about the origin, evolution, and presence of life in the universe. In the late 1950s, one of Sagan's mentors invited him to help the National Academy of Sciences with some "spadework, mainly consultation, on the generalities of biological probes." *Sputnik* had just launched, and the Soviets were hastily preparing landers for the moon. American scientists were beginning to worry about the secretive program, about whether the Russians were paying sufficient attention to things like sterilization, whether they might be risking humanity's chance to study life beyond Earth. NASA was already laying plans to protect the moon from contamination, and it seemed like a good time to begin thinking about other planets too.

Sagan's move was an auspicious one. It placed him for the first time in contact with the luminaries directing the American space program, at the beginning of what would ultimately become a watershed moment in Mars science: the shift from looking for life in photographs to looking for life in ways that couldn't possibly show up on film. It would require new instruments, new miniaturization, and, most important, actual scientific experimentation on the surface of the planet itself. But landing on a planet was very different from flying by it or orbiting around it. We could learn so much more, which inevitably meant that more was at stake, professionally and personally, for all involved.

The realignment of Mars science began as soon as the National Academies panel coalesced. The participants—including young Sagan as a PhD student—divided themselves into an East Coast group, EASTEX, and a West Coast group, WESTEX. The panel's first order of business was to set a path for their efforts, and at one of the earliest EASTEX meetings, one of the group members bombastically lamented that no devices had been invented that could be used for life detection. What was needed, he felt, was a simple contraption that would register "yes, there's life in this sample!" or "no, no life." It was a compelling, and ultimately transformative, call to arms.

The first scientist to propose such an experiment was a good-natured, gentle microbiologist named Wolf Vishniac. Vishniac was not the first in his family to explore the realm of the very small. His father, whom *The New Yorker* once described as "undoubtedly the world's leading photographer of microscopic life," had a nearly messianic view of his trade. "Nature, God, whatever you want to call the creator of the universe," the elder Vishniac had once opined, "comes through the microscope clearly and strongly. Everything made by human hands looks terrible under magnification—crude, rough, and unsymmetrical. But in nature, every bit of life is lovely. And the more magnification we use, the more details are brought out, perfectly formed, like endless sets of boxes within boxes."

Young Vishniac had grown up in Berlin, culturing algae at home, which he in turn fed to fairy shrimp, which he in turn fed to seahorses. He set sail for America in 1940 aboard an American Export Lines steamer with his family and eighty other Jewish refugees. Despite speaking little English, he landed a spot at Brooklyn College the following year, then went on to Stanford. As a young professor at Yale, he made a name for himself by working out some details of photosynthesis and investigating how microbes used sulfur as an energy source, before moving on to the University of Rochester.

Despite their rather divergent scientific proclivities, Vishniac and Sagan became dear friends. They had been trained very differently: Sagan, the astronomer, with his cameras and spectrometers, and Vishniac, the biologist, with his slides and test tubes. Vishniac loved to tinker and had a knack for engineering, whereas Sagan was butterfingered. And their personalities also set them apart. Vishniac had a quiet nature. While he could sometimes be found talking to his local Kiwanis chapter, members of the police department, and other luncheon groups, Sagan had the world as his pulpit. But each, in his own way, was startlingly imaginative, and each made profound contributions to Mars exploration.

IN 1959, AROUND the time Sagan was starting a postdoctoral fellowship at Berkeley, Vishniac received a grant of $4,485 for what he dubbed the "Wolf Trap," mocking his own first name. It was a life detection concept for Mars that was based on the idea that microbes would change the environment as they grew. Vishniac envisioned that after a soil sample was sucked up from the ground through a nozzle and dumped into nutrient-rich water, growing Martian microbes would produce changes in the culture media that could be measured, allowing scientists on Earth to see what was happening. A change in acidity could be picked up by a pH probe, and increased cloudiness—an indication of rapid growth—could be detected with optical sensors. Together the measurements would provide an independent check on each other. Exponential increases—a skyward swoop of the curve—would be particularly indicative of the proliferation of small microorganisms.

Within two years, Vishniac had a working model of the Wolf Trap, a huge scientific achievement and a shockingly quick response to the EASTEX challenge. Yet not everybody was impressed—including Vishniac's father-in-law. As a postdoc, Vishniac had fallen in love with and married Helen Simpson, the daughter of one of the most influential paleontologists and evolutionary biologists of the twentieth century. The elder Simpson had little regard for life detection. He'd publicly taunted biologists for agreeing that the first and foremost task in space science should be the search for alien life. He teased them about their "new science of extraterrestrial life, sometimes called *exobiology*," deeming it "a curious development in view of the fact that this 'science' has yet to demonstrate that its subject matter exists!"

But Vishniac persisted, and slowly he was joined by other microbiologists and biochemists who were thinking up new ways to miniaturize laboratory experiments to look for life on Mars. After the Wolf Trap, the next furthest along was an instrument dubbed "Gulliver"—an homage to Jonathan Swift, aptly named in its quest for Lilliputian life. Designed by a sanitation engineer, Gulliver sought to capitalize on one of the most common ways to detect microbes in swimming pools, oceans, and drinking water, particularly contaminants like fecal coliform. The idea was simply to monitor a culture for bubbles of carbon dioxide using a carbon-14 tracer. Some of the early designs envisioned tiny harpoons, fired like mortars from the base of the lander, trailed by seven and a half meters of kite line. The kite line would be coated in silicone grease to stick to soil particles, making them easy to reel back into the instrument to analyze.

By the early 1960s, NASA was funding nearly twenty life detection concepts, but notably, none of these instruments—not even Vishniac's, the son of a microscope evangelist—were designed to take images. It was too complicated to prepare specimens and search glass slides for growth, and transmitting the images required too much data. Microscopy would have to go, and with that jettisoning, the search for life crossed a major threshold: For the first time, life would not be something you saw; it would be something you measured in an interplanetary laboratory.

WHEN THE *MARINER 4* mission began releasing findings in 1965, the new tribe of exobiologists was as stunned as the rest of the world. The extreme aridity, the extreme cold, the extremely low atmospheric pressure all raised serious doubts about how life could survive on Mars, and suddenly it seemed like they might be wasting their time. Until the exobiologists could articulate a theory for survival in such an inhospitable place, their hard work would look like a fool's errand.

Naturally, it was Sagan and Vishniac who stepped to the plate. In response to the cratered images, Vishniac penned a moving letter to the Senate chairman of the Committee on Aeronautical and Space Sciences, arguing that craters might even contribute to "a diversification of the environment, with the creation of ecological niches favorable to colonization by living organisms"—that they might enable entry to deeper geological formations, protect organisms from radiation "in the shadows," and provide a chance for essential elements to gather.

Meanwhile, Sagan scrutinized hundreds of photographs of Earth, trying to show that the *Mariner* results didn't necessarily preclude life. He looked first to weather satellites: to TIROS-1 and Nimbus, which were just opening the door to space-based remote sensing. At one-thousand-meter resolution, he could see no roads or buildings, no rectilinear patterns, no life whatsoever over New York City, Moscow, Paris, or Peking, he pronounced. He then amassed eighteen hundred images snapped of Earth by the Apollo and Gemini astronauts, with ten times greater resolution. At one hundred meters a pixel, only a handful of the pictures showed even a trace of anything human. And even then, he argued, you had to know what to look for: A tilled field or a thin track of road would mean nothing to an observer who was unfamiliar with life on Earth. To such a newcomer, our human workings would be invisible, and the eighteen hundred images might lead to the erroneous conclusion that Earth was devoid of life. Our sense that we'd left an imprint, that we'd had a profound influence on the physical world, it was all wrong. So too was our assumption that if life existed on Mars, we would have spotted it by now: After all, if we were undetectable from above, Martian life might be as well.

Although Sagan wasn't a biologist—his focus was on spectroscopy and imaging—he'd begun dabbling in the life sciences. Building on little-known experiments that had been conducted near San Antonio, Sagan began constructing "Mars jars," which were small chambers simulating the inhospitable Martian surface and atmosphere. He filled the jars with terrestrial microbes, in temperatures ranging from minus 80 degrees Celsius at midnight to freezing by noon, under harsh ultraviolet light. What he discovered struck him as remarkable. As he wrote of the work, "There were always a fair number of varieties of terrestrial microbes that did not need oxygen; that temporarily closed up shop when the temperatures dropped too low; that hid from the ultraviolet light under pebbles or thin layers of sand." In experiments where tiny amounts of water were present, he was delighted to find that the microbes actually grew. "If terrestrial microbes can survive the Martian environment," he concluded, "how much better Martian microbes, if they exist, must do on Mars."

At the same time, NASA embarked on its own effort to assess whether life could survive in harsh, cold conditions. It sent scientists down to the McMurdo Dry Valleys of Antarctica—the small sliver of the great white continent that is ice-free—to collect samples from one of the more Mars-like places on Earth and distributed samples to the researchers designing life detection instruments for calibration and testing. As part of the effort, a scientist named Norman Horowitz, said to resemble a fox terrier, tried every which way to culture life from the samples, but not a single bacterium grew under any of the treatment conditions. Horowitz published a *Science* article in 1969, swinging the ax. While the cold and barren Dry Valleys contained significant amounts of organic carbon molecules—the building blocks of life—vast tracts of land were in fact sterile. And if life couldn't even survive in Antarctica, how could it survive on Mars?

This finding alarmed Horowitz, who was already deeply tied to the exobiology enterprise. In response, he led a last-minute effort to develop a different instrument concept. It co-opted the exact same detection principle as Gulliver, but whereas Gulliver moistened the soil with water and the Wolf Trap drenched it, Horowitz's experiment was completely dry. There was to be a small lightbulb in the testing chamber, and if organisms similar to simple algae in the soil were able to pull any of the carbon-14 tracer from the chamber's air into their bodies, it would be evident once the chamber was flushed and the sample baked.

Soon Horowitz's new experiment was selected as one of four instruments to fly on the *Viking* mission—alongside a modified version of Gulliver, the Wolf Trap, and a final instrument nicknamed the "chicken soup," run by Vance Oyama. This last one was based on the idea that the more food the Martian microbes got, the more respiration there would be to measure. Oyama's instrument added lots of water and a wide variety of nutrients—the chicken soup—and then watched for changes in the composition of gases in the chamber as a result of metabolic activity. But by the time the instrument selections were made, the Dry Valleys results had convinced Horowitz that the chance of life on Mars was negligible—so low, in fact, that he even argued that it was pointless to sterilize the spacecraft. His instrument and the other three were exceedingly likely to fail.

INCREASINGLY, THE ACADEMIC community was looking askance at the entire effort. EASTEX and WESTEX had drawn a number of scientific heavyweights, but they were mostly scientists who were far along in their careers, already decorated with awards and endowed chairs. Exobiology was a safe side game for Nobel Prize winners. But it was risky as hell to stake one's career on it—something Sagan and Vishniac found out the hard way.

After landing an assistant professor position at Harvard in the early 1960s, Sagan had little reason to doubt his tenure prospects, even though the bar was high. He'd conducted peer-reviewed research for years, and not just in exobiology. His pioneering work on the makeup of the Venusian atmosphere—and the idea that its incredibly hot temperatures were due to the greenhouse effect—had been proved largely correct by the data returned from the *Mariner 2* mission to Venus in 1962. He'd written scores of scientific articles and had received NASA funding for his work. But in his fifth year, just as *Viking* was beginning to be planned, he was told point-blank that tenure would not be forthcoming. Little more was communicated.

Frantic, he approached crosstown-rival MIT about the possibilities there, but after an initial warm response from the geology and geophysics department, their interest suddenly and inexplicably cooled. Sagan knew exobiology was precarious. Vishniac's father-in-law was on the faculty at Harvard, and many of Sagan's fellow astronomers shared the elder scientist's skeptical views. As they collected their coffee and paced the hallways, they quipped that exobiology's "speculations cannot be confirmed with observations or experiments and so it is not a science; it has no data. It only sounds like a science." But Sagan had his supporters; his résumé was sterling and his funding secure, so what was happening?

Unbeknownst to Sagan, his star had dimmed because he had been abandoned by one of his oldest, most trusted mentors. His undergraduate thesis advisor from the University of Chicago, Harold Urey, had written a tenure letter to Harvard, and a similar one to MIT, denigrating Sagan's work. He described "the sort of activity that Carl Sagan has been engaged in for years" as "very long, wordy, voluminous papers that have comparatively little value . . . many, many words, oftentimes quite useless." He characterized Sagan as a kind of planetary dilettante who had "dashed all over the field of the planets—life, origin of life, atmospheres, all sorts of things." Hearkening back to his experience as Sagan's mentor at University of Chicago, he twisted the knife: "Personally I mistrust[ed] his work from the beginning. He is a smart fellow and he is interesting to talk to. Perhaps he will be a valuable professor at your institution. But for years I have been disturbed by this sort of thing . . ."

Sagan had no idea Urey considered him a dilettante, and he wouldn't find out for years—not until long after he finally secured a post at Cornell. The news would come out of the blue in the form of an apology that arrived in the post. "I have been completely wrong," Urey wrote on September 17, 1973, as he asked for Sagan's forgiveness and friendship. "I admire the things you do and the vigor with which you attack them." Sagan responded magnanimously, but what cold comfort it must have given Sagan to know that his trusted mentor regretted having secretly undermined his career.

With the benefit of hindsight, it seems that Urey was cruelly slow to see the value of Sagan's scientific work. Even if some of his papers meandered, Sagan was a tireless worker whose research made important contributions. Urey should have been the first, not the last, to realize this: After all, when Sagan received Urey's biting criticisms of his undergraduate draft thesis, he carefully redid the entire thing. As a PhD student, years before *Mariner 9,* Sagan had bravely taken on his own PhD advisor about the wave of darkening, not by discounting data but by embracing it: He argued correctly that the known evidence better supported a lifting and settling of dust. And Sagan's PhD advisor was no ordinary scientist but Gerard Kuiper, who at the time was the most prominent planetary scientist in the world. That Urey would not have understood the depth of his commitment to science, or his talent, must have been crushing to Sagan, even all those years later. Such were the perils of championing a new scientific discipline and of doing it so publicly.

A bright spot was that Sagan's position at Cornell placed him not far from Vishniac, who was at the University of Rochester. They would see each other often, not only in New York State but also, as meetings took them to the far corners of the world, in places like Tokyo, Barcelona, Leningrad, and Konstanz. It was a friendship between kindred spirits. And then of course there was the excitement of *Viking*'s life detection efforts. Although they recognized the scientific value of a negative result—and appreciated that it would place at humanity's disposal a planetary surface that hadn't been "turned topsy-turvy" by living organisms, a kind of control group—they both dearly hoped that *Viking* would be successful, that life would at least be hinted at if not definitively discovered.

The mission's engineering task, however, was enormous. By the early 1970s, *Viking* was millions of dollars over budget, with time running out. The biology package, which had been billed as the "greatest experiment in the history of science," was the most behind. The automation was extremely challenging: There were forty thousand parts, half of them transistors that had to be assembled, plus tiny ovens, nutrient-containing ampules that had to be broken on command, bottled radioactive gases, Geiger counters, and a xenon lamp to mimic the light of the sun. The instruments and the receiving end of a sample delivery system had to be shoehorned into a sixteen-kilogram box, measuring about the size of a milk crate. Somewhere had to fit reservoirs for helium, krypton, and carbon dioxide, fifteen meters of stainless-steel tubing, heaters, coolers, test cells, dump cells, a thermostat, and a carousel.

Bad news came quickly for Vishniac. He had originally been appointed to lead the biology team, but unable to keep up with the draconian deadlines—Vishniac tended to "let everyone have his say"—he was replaced by someone with a considerably more authoritative air. Then, with little warning, *Viking* dropped the Wolf Trap altogether. The original estimated cost of $13.7 million for building the life detection experiments package had ballooned to more than $59 million. It became clear that the biology payload would have to be simplified and that at least one instrument had to be cut. The one the team chose was the light-scattering experiment, Vishniac's brainchild. Cloudiness might result from the dispersion of fine soil particles, not just microbial proliferation, and whereas all of the other instruments could detect resting metabolisms, the Wolf Trap required conditions of growth. It was too complicated and already running behind schedule. In the background was Horowitz—persuasively arguing that the Wolf Trap, like the chicken soup, would probably drown any Martian life.

When he found out the news, Vishniac was devastated. For more than a decade, his scientific career had been focused on life detection on Mars and the *Viking* mission, on what many of his peers considered a boondoggle. And now he would not have the chance to prove his critics wrong. He remained on the *Viking* team, but he no longer received NASA money to support his research, and he struggled to secure any funds from the National Institutes of Health and the National Science Foundation. "The consequences of my change in status in the *Viking* team have been far-reaching, as you know, not to say disastrous," he told the mission's project scientist. "It is essential that I recapture some sort of standing in the academic world."

With time on his hands in 1972, and then again in 1973, Vishniac decided to venture down to Antarctica himself. He was nothing if not determined. He had a weak arm from a birth injury, with limited mobility, but nevertheless he'd taken up winter sports as a teenager. He had a stutter, but he'd become a public lecturer and professor.

He'd even failed the glucose tolerance test as part of a Navy physical to travel to Antarctica, but "knowing something about the chemistry of the test," he'd devised a way to pass it on his second try. He was intent on showing that Horowitz was wrong, that life could exist in extremely arid conditions, and that the *Viking* mission wasn't, in fact, completely hopeless.

At the beginning of his expedition in 1973, Vishniac gingerly tucked glass slides coated with nutrients into folds of soil high in the Asgard Range, named, appropriately enough, after the homeland of the Viking gods. The experiment was like Sagan's Mars jars, but it was conducted in nature, in one of the most Mars-like places on Earth.

A month later, two weeks before Christmas, Vishniac began collecting his slides. It was nearing midnight as he made the rounds, although the sun never sets in Antarctica in December. His colleague Zeddie Bowen had stayed at their camp to await a supply plane. The men often ventured out alone: "Always in good weather and with an expected route and timetable. The route was not dangerous." But twelve hours later, when Vishniac hadn't returned, Bowen went looking for him. At worst, he envisioned a broken ankle. "What I really expected was to find him distracted by some fascinating new discovery or observation." Instead, he found Vishniac's lifeless body at the base of a cliff, beneath a hundred-and-fifty-meter ice slope. He'd taken a different route, "following his curiosity," then a wrong step. His body was recovered by the crew of a Navy helicopter, then transferred home to Rochester. Vishniac left behind his wife, Helen, their two teenage boys, and a devastated *Viking* biology team, who respected Vishniac deeply, even though they'd been forced to jettison his instrument. They gathered at the funeral, held in the cold New York winter.

WITH COST OVERRUNS and engineering challenges, and the terrible tragedy of Vishniac's death, it was a near miracle that the *Viking* biology package was delivered in time for launch. The absence of the Wolf Trap was ever so palpable, and the anguished team struggled to complete their work. The time pressures on Viking were such that, had the Wolf Trap remained part of the mission, Vishniac probably wouldn't have been in Antarctica; he would have been there with them, finishing the project.

But finally, two biology payloads, one for each lander, were complete. On March 7, 1975, NASA's Langley Research Center wrote to the *Viking* project office and contractors announcing that the streamlined *Viking* biology suite was at long last in its shipping box, ready for delivery.

By the time the instruments arrived in Florida, there were nearly daily thunderstorms at the cape, and time was running out. Even though Horowitz had argued that fears about terrestrial microorganisms multiplying and contaminating the planet were silly, the landers were nevertheless sealed beneath their ablation shields and inched like giant mushrooms into a 112-degree-Celsius oven to be sterilized. There they steeped beneath searing clouds of nitrogen gas for forty hours. Assembly and testing continued, right up until August 20, 1975, when *Viking 1,* the first of the identical spacecraft, soared into space on a Titan IIIE rocket. *Viking 2* followed just three weeks later.

WITH GREAT FANFARE, the first lander was scheduled to touch down on the surface of Mars on America's bicentennial. The nominal landing site was near a deep and wide swath of terrain at the mouth of Valles Marineris, a floodplain where water was thought to have flowed out of the giant canyon. There in Chryse Planitia, the "plains of gold," was the confluence of several ancient river channels. The site was also far below the mean surface of the planet, hinting at the possibility that snow, ice, and perhaps even traces of transient liquid water—less likely to evaporate under the greater atmospheric pressure—might be present.

On the other side of the planet was the nominal location for the landing site of *Viking 2*. Cydonia, right at the lower edge of the north polar hood, was near the ice. As a result, the atmospheric humidity was anticipated to be high, and it was hypothesized that microbes might even drink water from passing clouds.

Both sites had been chosen for their smoothness in the *Mariner 9* pictures, with landing zones carefully fitted between craters and canyons, but as the first orbital color images began trickling back from Mars in June of 1976, the terrain was nothing like the team expected. The orbiters had been designed to photograph the planet to help assess the landing sites before the landers detached, but rocky knobs, steep slopes, and hidden small craters suddenly spattered into focus. The team realized that even after the giant dust storm of 1971 had subsided, a haze must have persisted in the air, filtering the *Mariner 9* images, dampening the contrast, making the terrain appear softer and more muted than it really was.

"It may be that we don't understand Mars at all," wrote the mission's project manager, sending jitters throughout JPL. "But we shall find a place to land, I think . . ." It was generally agreed that crashing into Mars on the bicentennial would be a bad idea, so the July 4 date was scrubbed, and the team began scrambling to find an alternative. Again, time was short. The landings had been scheduled more or less on top of each other, and the communications network would soon become strained.

So for the next two weeks, the orbiter went tacking across the surface, searching for safe ground. The landing site had to be a low-lying area so that the lander's huge parachute—sixteen meters wide, as large as the team could make it—would be able to catch enough atmosphere to slow down the speeding capsule. The site couldn't be out of communication range of the orbiters or too cold for the instruments to operate, which eliminated large swaths of the high latitudes. It had to be fairly flat, or else the lander might land at an angle, leaving the mechanical arms "waving helplessly" above the surface. But hard lava had to be avoided too, as there would be no soil for the arm to collect and analyze.

Finally, a site was chosen for the first lander. It was also in Chryse, though not near the confluence of the ancient river channels. On July 20, *Viking 1* split in two. Like an artillery shell, the descent capsule barreled toward the surface of Mars on a ballistic trajectory, while the orbiter continued to circle. At nine hundred kilometers an hour, about six kilometers above the surface, the parachute deployed. A kilometer and a half above the surface, after the aeroshell had been jettisoned and the legs had begun to unfold, the parachute was cut and the retrorockets were activated. Within seconds, the beetle-shaped lander came to rest softly on the surface. Cheers erupted throughout JPL. The sheer fact that a spacecraft could land on another planet was itself extraordinary, considering it had largely been accomplished with just vellum and notepads.

The images of the landing site from the *Viking* orbiter were a vast improvement over the *Mariner 9* images, which from fifteen hundred kilometers up hadn't been able to spot anything smaller than the Rose Bowl. Even the *Viking* orbiters, with their better coverage, could only resolve features bigger than about a hundred meters across. Because no one knew what the surface of Mars would actually be like, the first image was of the lander's foot, simply to make sure the surface was solid.

The next image was of rocky terrain beneath a bright blue sky. Upon seeing it, one scientist began wandering down the halls of JPL, cheerily singing, "Blue skies, do dah dah dah . . ." Many on the team, Sagan included, had predicted the sky would be black overhead because of the very thin atmosphere, then a lighter blue-black near the horizon, where there was more atmosphere to look through. Strange that the sky was so bright.

The image-processing laboratory, which corrected for things like the sun angle, uneven shading, and curvature distortions, slowly began to realize that the first image had registered the appearance of the atmosphere incorrectly. The color on the lander's facsimile cameras had to be calibrated, numerically re-creating the hues because the color diodes were also sensitive to infrared light. The engineers soon discovered that the sky on Mars wasn't a luminous bright blue, but, weirdly, it wasn't a dark blue-black either. It was full of light and orange, the color of butterscotch—light reflecting off billions of tiny dust particles in the air.

The image was quickly corrected, and as additional images came down, Sagan eagerly studied them, buoyed by optimism, acutely aware that the cameras were the only instruments, in principle, that could prove the existence of life on Mars in a single observation. He had lamented that the search for landing sites would plop the *Viking* probes down on the most boring parts on Mars: "We knew we had chosen dull places. But we could hope." In the end, the most interesting feature to photograph was "Big Joe," a boulder a few meters out of reach. In a press conference on the mission's eleventh day, Sagan joked with the gathered reporters that no rocks had gotten up and moved away, at least not yet.

Despite the stillness of the terrain, the biology experiments suggested that something very exciting was happening in Chryse Planitia. Initial samples had been scooped up with a telescoping arm from a bare patch of soil in front of a rock named "Shadow." After being delivered to little buckets "like hoppers on an electric train," they slowly rode inside the lander to be analyzed by the three life detection instruments as well as two others assessing the sample's chemistry and mineralogy. The team was prepared to wait days or weeks for the incubations to yield results. Yet, miraculously, Oyama's chicken soup experiment kicked up gases in just a couple of hours, potentially indicating a type of wildly fast metabolism. A spew of radioactively labeled carbon dioxide also was detected, by the modified Gulliver. The team was ecstatic. "We were so excited, we sent out and got champagne, cigars," the instrument's lead recalled. Then he and his associates solemnly sat down to sign and certify the data printouts, aware of the magnitude of what of they were doing. In his mind, the instruments had satisfied the mission requirements for the detection of life.

But the experiments soon went haywire. It was as if the instruments picked up too much life, then none at all: The readings flashed and then died away. The team realized that the rate of responses registered was faster than even the most fertile soils on Earth and that the chicken soup experiment had produced a rapid surge of gas even before the nutrients had been added. They started to wonder if the water introduced in those experiments might be causing a chain of powerful chemical reactions. Perhaps instead of life, there was something chemically corrosive in the soil?

The nail in the coffin was the gas chromatograph-mass spectrometer, the chemistry instrument designed to detect carbon-based organic molecules. The results had been delayed by several days because the scoop had gotten jammed. But when the experiment was finally run, it found no organics whatsoever. Even the lifeless moon contained *some* organics: simple molecules that had rained down from space on comets and meteorites.

Weeks later, similar results trickled back from Utopia Planitia, the landing site for *Viking 2*, some five thousand kilometers away. The biology team tried everything they could think of: shorter experiments, longer ones, different combinations. In the end, nearly everyone concluded that those initial detections must have been false positives. How could there be life without any organic molecules, without any of life's building blocks? Horowitz declared that it was "virtually certain" that Earth was the only life-bearing planet in our region of the galaxy, now that we'd found Mars to be utterly barren: "We have awakened from a dream . . ." In his view, *Viking* had found not only no life on Mars but also why there could be no life: The planet was devoid of water and suffused with cosmic galactic rays, both of which were sufficiently sterilizing. He concluded that oxidants like hydrogen peroxide were laced throughout the soils, the result of billions of years of intense radiation. As a result, there were corrosive free radicals everywhere, roving reactive atoms and molecules with unpaired electrons—so many that complex chemistry was constantly under attack.

Predictably, some in the scientific community turned their frustration on Sagan, criticizing his ludicrous optimism, arguing that he had only set the public up for disappointment. Sagan had playfully warned reporters about the possibility that the experiments might come up empty while creatures were "placidly munching on the zirconium paint on the outside of the lander." He had even suggested that the mission add cameras, bait, and a lighting system to the lander to lure Martian life-forms to the craft, having gone so far as to run tests in the Great Sand Dunes National Monument using a snake, two tortoises, and a chameleon. But of course there were no silicon-based giraffes—and how very irresponsible to pretend there could be. In Sagan they saw a showman, a huckster, a megastar. And it rankled them.

. . .

FOR HIS PART, Sagan still combed through many of the thousands of orbital images that *Viking* had collected for signs of life. He and his students flipped through image after image, quadrant after quadrant, finding nothing of note. The mission results had sobered him, along with most every other Mars scientist. There was little doubt that "the greatest experiment in the history of science" had failed, and not even Sagan could object when the *Viking* team stopped the single-slit scanners it was using to detect motion on the surface, one of the instruments he had fought the hardest to include. The *Viking* data, such as it was, would be all the Mars community would have to work with for the next twenty years, for it would be that long before a spacecraft returned to the planet. Exobiology had flared like a match. And then burned out.

CHAPTER 5

Stone from the Sky

OUT IN THE frozen terrain of Antarctica, where no rain has fallen in two million years—in the land of bleached skies, of no dogs or children—Wolf Vishniac had been attempting to connect what was known about the geology of Mars to what he knew about biology on Earth, to understand whether microbes could survive the harsh conditions. It's of little consolation, but before Vishniac died, he knew he'd found what he was looking for: life in the "lifeless" soils. The microscope slides he'd pulled from the ground of the Asgard Range showed stunning constellations of growth. When held to the light, they looked like small, shining galaxies.

The cells had taken to the slides, but not to his colleagues' petri dishes, because Vishniac had basically left the microbes alone, allowing them to grow in their natural environment. After his death, they were sent back to his wife, Helen, who found that they contained hundreds of cells, and not only microbes but also complex eukaryotes.

Among Vishniac's personal effects was also a bag of cold desert sandstones. He had written across it "Samples for Imre Friedmann."

Friedmann was a fellow microbiologist, at Florida State University. For ages, he had wanted to see if life could survive inside rocks, but he hadn't succeeded in securing funding to go to Antarctica. He'd asked Vishniac to pick up some rocks for him, and when he received the bag from Helen, he made a groundbreaking discovery. In 1976, Friedmann published his results in *Science:* Vibrant unicellular blue-green algae had colonized the air spaces inside the porous sandstone, using tiny rock houses as protection against the elements. Life, in other words, could live not only in freezing-cold deserts but also in freezing-cold rocks.

These discoveries—and others—kicked off a new phase in the search for life after the final, feeble radio wave from *Viking 1* oscillated to Earth from the western slope of Chryse Planitia in 1982. Some of the starry-eyed graduate students who wrote their dissertations on Mars in the 1970s—the young scientists who were left with barely any data to work with—turned their attention to our own planet in the 1980s and 1990s. A new field of extremophile biology was born: investigating the crooks and crevices of our planet to better understand the limits of life.

Microbes were soon discovered in brine pools many times the saltiness of seawater, even supersaturated with methane, and in lakes with the pH of Drano. Scientists who began exploring the eternal darkness of the deep sea found that there was not only life but in fact a rich and intricate ecosystem. Despite toxic sulfide gas and temperatures hot enough to melt lead, hydrothermal vents were teeming microbial communities. There were thickets of tube worms, some more than two meters long, waving like human arms. They were tipped with feathery red plumes, seen for the first time as the lights of submersibles cut across the seafloor. They were surviving at pressures that no one imagined life could endure, far beyond the reach of the sun's photons. A new type of metabolism had to be powering this world, one that wasn't using photosynthesis as a source of energy.

Microbes were found laced throughout Yellowstone's evaporites—soft, salt-crusted, sedimentary rocks that formed from the evaporation of waters in strange bubbling pools. In Octopus Spring, there were pink hair-like strands of *Thermus aquaticus,* tiny organisms that could live and reproduce at extremely high temperatures. *Pseudomonas bathycetes* was pulled from the crushing pressures of the Mariana Trench, and *Deinococcus radiodurans* was scraped out of the waste of nuclear reactors. Life, it seemed, was everywhere.

THE STAGE WAS set for the discovery of a little rock shaped like a potato in the Allan Hills of Antarctica. In 1984, two days after Christmas, a young scientist named Robbie Score spotted a small dark spot on an ice sheet more than two hundred kilometers south of the Asgard Range. She was snowmobiling with a team of researchers from the Antarctic Search for Meteorites. She sped over on her Ski-Doo to have a closer look, then signaled to her colleagues. It looked almost green against the metallic whiteness. The team extensively photographed it, gingerly placed it in a clear plastic bag, and labeled it "ALH84001." ALH for Allan Hills; 84 for 1984; and 001 because it was the first find of the year.

The Antarctic Search for Meteorites was established because more meteorites are found in Antarctica than anywhere else. It's not that more fall there, just that they are easier to see. In fact, in certain parts of Antarctica, most of the rocks you find are meteorites. The slowly flowing ice gathers them in the immense interior of the continent. Glaciers creep downhill until they fall into the sea or run into a mountain range. If blocked, the trapped ice ablates away, which brings the frozen meteorites back to the surface. On the flanks of the Transantarctic Mountains, which form a spine down the continent, meteorites accumulate at concentrations many times higher than anywhere on Earth. They're easy to spot, like flecks of pepper on smooth white porcelain.

At the end of the Antarctic summer, ALH84001 traveled back to the United States in a climate-controlled shipping container with all the others found that season. When the rock was brought to a cleanroom in Houston in early 1985, a half-gram chip was cleaved and sent off for classification at the Smithsonian National Museum of Natural History. The young curator who did the analysis classified the meteorite as a diogenite, most likely from Vesta, a large asteroid in the asteroid belt between Mars and Jupiter. There were some strange patches of brown iron-rich carbonate that were unusual for Vesta, but the curator assumed they were due to weathering processes here on Earth.

For seven years, ALH84001 sat in a tightly secured vault at the Johnson Space Center. Then in 1992 a puzzled researcher wandered down the hall of nearby Building 31. He'd been doing a systematic study of fragments thought to have arrived on Earth from Vesta, but there was one rock that he simply couldn't place: the diogenite with the carbonate. He stopped by the office of David McKay.

McKay was a tall geologist with stooped posture and a hasty gait. He wore wire glasses over his rounded cheeks and had an easygoing demeanor. Originally from Titusville, Pennsylvania, he had moved to Tulsa, Oklahoma, in the sixth grade—his father had an accounting job with the Kewanee Oil Company—and then to Houston, Texas. He rambled for a while, working on remote offshore oil rigs and doing solitary surveying work in the desert, before returning to Rice University in Houston, his alma mater, for a PhD in geology. He'd been sitting in the Rice football stadium in 1962 when John F. Kennedy announced that the United States would go to the moon before the end of the decade. "Why choose this as our goal?" Kennedy asked. "We choose to go to the moon in this decade and do the other things, not because they are easy, but because they are hard, because that goal will serve to organize and measure the best of our energies and skills, because that challenge is one that we are willing to accept, one we are unwilling to postpone, and one which we intend to win." Inspired, McKay went on to secure a position at Johnson Space Center, a newly formed complex that was quickly overtaking the salt grass south of Houston. He'd been there ever since.

As McKay examined ALH84001 in the early 1990s, expecting it to be from Vesta, he couldn't help wondering about the rock's origin, just as centuries of scientists had pondered the origin of meteorites they'd found. The mere idea that rocks could plunge from the sky had once drawn ridicule. A famous eighteenth-century mineralogist remarked that "in our time it would be unpardonable to consider such fairy tales even probable." Some believed the bizarre objects were volcanic rocks, lofted like small bombs during an eruption, or rocks that had condensed in hail-filled clouds, or rocks that had been hit by lightning, giving rise to the name "thunderstones." Isaac Newton's work, which suggested that no small objects would exist in interplanetary space, wasn't questioned until the turn of the nineteenth century, when a German physicist first suggested, to great mockery, that meteorites from space caused fireballs and might themselves be "world fragments."

McKay wondered if the rock might in fact be a kind of meteorite called SNC, or "snick"—the shergottite, nakhlites, and chassignites—named for three witnessed falls near the villages of Shergotty in India in 1865, El-Nakhla in Egypt in 1911, and Chassigny in France in 1815. Loud sonic booms accompanied all three. A piece of the first nakhlite was said to have landed on a dog. It was clear that the three rocks from those tiny villages had curious properties, ones that set them apart from all other rocks. But where were they from?

As the group of three SNCs grew over the years, the mystery of their origin intensified, until it was discovered in 1983 that there were gas vesicles in one of the rocks, holding tiny beads of atmosphere. That there was any air at all trapped in the meteorites was remarkable, ruling out all airless worlds, including comets, asteroids, the moon, and Mercury, but leaving open the possibility that a planet with an atmosphere like Mars might be the source. Then everything about the chemical signatures in the atmosphere began to line up. As the vesicles were pierced, the gas matched perfectly with the ratios determined for the Martian atmosphere both by Earth-based spectroscopy as well as the direct measurements made by the *Viking* landers. In addition, new models emerged to explain how fragments of material, called "spalls," could be ejected from the surface of Mars without being melted—or, for that matter, completely vaporized.

In some ways ALH84001 looked to Dave McKay like the other Mars meteorites, but at the same time it was clearly different. For one, it was three times older. When one of McKay's colleagues ran an analysis to determine how long the rock had been exposed to cosmic rays—the radiation that is constantly bombarding the surface of the planets from space—he got an astonishing number: At the time, it appeared to have formed only fifty million years after the birth of our solar system, making ALH84001 the oldest rock from any planet ever discovered, including Earth. An impact sixteen million years ago likely cleaved it from the subsurface of Mars, where it had been protected from the harsh surface environment, then flung it into space on a vector aimed at Earth. A set of exposure age calculations put the date of ALH84001's arrival in Antarctica at about thirteen thousand years ago—before the beginning of agriculture, before the rise of civilization. ALH84001 landed just as the last ice age was ending, as glaciers began pulling back from their vast hold on our planet. The fistful of rock was frozen beneath the ground for all those years, cocooned from wind, storms, and sunlight.

McKay liked to listen to Enya, an Irish singer with an otherworldly voice, while he worked in the lab, often late into the night. As he peered into the rock, he began noticing orangey knobs of carbonate, the first of several strange discoveries. The knobs were extremely bizarre: ringed like the eyes of owls. The proportion of carbonates in the meteorite, about 1 percent, was far higher than anyone would ever expect to find in a rock that cooled from a hot volcanic mass. Carbonates on Earth, like the vast sheets of limestone that cover North America, almost all form in the presence of water and in the temperature range where that water is in a liquid form. The presence of carbonates in the rock suggested Mars had been awash in water, that the rock had formed in a habitable environment.

Soon, delicate magnetite crystals laced throughout the carbonates like strings of beads were spotted by McKay's colleague Kathie Thomas-Keprta. This was another unexpected finding. On Earth, microbes produce these crystalline arrangements, which serve as tiny compasses for the microbes as they glide around. The magnetite crystals in ALH84001 were also extraordinarily pure. Magnetite formed by indiscriminate geologic processes typically contains magnesium, calcium, and iron, whereas microbes tend to select only for magnetite with iron, which has the best magnetic properties. Under natural conditions, magnetic minerals form under different pH conditions from carbonates, so the presence of the carbonates and magnetite crystals together was unusual and potentially indicative of life. Billions of years ago, did microbes float through the Martian seas, guided by the tug of an ancient magnetic field?

If so, McKay and his colleagues reasoned, there should be remnants of organic material, the building blocks of life, along with the magnetic structures. To test this idea, McKay sent samples of the meteorite to a widely respected laser chemist at Stanford University. In a matter of weeks, ringed clusters of carbon and hydrogen atoms called polycyclic aromatic hydrocarbons (PAHs) were detected. PAHs are found in oil, coal, and tar, in the charred remains of burned forests, and in the black residue of steaks grilled on an open flame.

McKay and his team still wavered. The PAHs weren't a smoking gun—while PAHs are commonly formed as a by-product of cellular decay on Earth, they are also formed by the creation of new stars. Even so, McKay seemed to have stumbled upon the first true evidence of organics on Mars. At the very least, the PAHs signified that a kind of chemistry conducive to life was at one point present on Mars. The PAHs were much more plentiful in the core of the rock, arguing against contamination, and not only that, the PAHs in ALH84001 appeared to cluster precisely where the carbonates and magnetic crystals were concentrated.

McKay continued to probe ALH84001 back at Johnson Space Center, and in a stroke of fortune, he was given access to a sophisticated new scanning electron microscope, one NASA had recently purchased to inspect space hardware for tiny cracks and flaws. McKay knew it would allow him to peer into the mineral structures of the meteorite at unprecedented resolution.

One day in January 1996, he and a colleague carefully placed a tiny speck of the meteorite into the machine and turned on the beam. As they looked down on what appeared to be entire continents of unfamiliar terrain, something bizarre suddenly came into view. Both of them sat completely still. There on the edge of one of the orange carbonate blobs was something that looked like a worm ascending a hill, suspended mid-crawl as if captured in some small Martian Pompeii. It was shaped like a rope, just fifty or a hundred nanometers across, and, wildly, it seemed to be segmented like a primitive microorganism. It appeared for all the world to be a nanobacteria fossil, *an actual fossil of life*.

McKay was so excited that he printed out the image and left it for his thirteen-year-old daughter to find. "What does that look like to you?" he asked her, trying to be casual, after she saw it. "Bacteria," she replied candidly.

The images solidified McKay's decision to publish the results. The carbonates, the magnetic minerals, and the PAHs appeared in every last sand-sized sample of the rock he and his team analyzed. All three features were suggestive of life. And those structures McKay had imaged could well be the visual remnants! There were other plausible explanations for each feature, but taken together, particularly in light of how they clustered in the same place, McKay concluded something astounding: that they had found the first ever evidence of primitive life on Mars.

He and his team carefully finished the manuscript and submitted it to *Science,* which lined up a slate of nine reviewers, including Carl Sagan, to weigh in on whether the article should be accepted for publication. In 1994, just two years earlier, Sagan had written an article himself noting that no microbes had been detected in rocks from the sky, at least not yet. He was dumbfounded when a copy of McKay's manuscript landed on his desk. With ALH84001, it seemed that his life's work had finally come to fruition, and just in time, as Sagan was battling bone marrow disease.

As soon as the paper was accepted, the journal scheduled a release date for the middle of August. Foreseeing the press frenzy, NASA went into lockdown, determined to keep the news under wraps until the paper's official publication date. That left McKay about two weeks to catch his breath before he would become one of the most famous scientists in history.

He and his family set off on a camping trip along the Frio River. The state park where they set up was located on soft rock that had formed a hundred million years ago in the margins of shifting seas. It was a geological wonderland. Prehistoric animals had left their footprints in the sand, and a plateau of limestone pushed upward along a curving fault. They drove below the canopies of cypress trees, slowly winding their way over the Cretaceous rocks.

LIKE MCKAY, I have always felt at home among rocks. When I was a child, my father loved dragging my sister and me out of the car to look at roadcuts, those sections of highway where the land had been sliced away to make room for the asphalt. There were roadcuts all along the Mountain Parkway, which is how we got from my house to my grandmother's in eastern Kentucky. My father's mother was a tough woman, and only got tougher as she aged. She stopped taking X-rays after my grandfather passed away, but she still gave porch perms, and she still made soup beans with rowdy irreverence in her dark, wood-paneled kitchen. My father would stop often along the way to search for fossils on those rock walls: the bryozoans and brachiopods, the roly-poly trilobites and ostracods, and the ancient crinoids, saltwater animals once attached by a stalk to the seafloor.

About halfway to Hazard, there is a stretch of parkway where the years skitter by in the tens of millions. The rocks there are stacked like pancakes, getting younger and younger with each passing kilometer. They chronicle the history of life as it climbed up onto the barren continents—as amphibians took hold, then insects, then sail-backed reptiles.

We'd be driving that road, my sister and I in the backseat of the car, when my father would open his thermos. We couldn't stand the smell of coffee. We'd roll and moan in protest, gasping for air from the open windows. As we pretended to suffocate, my father would seize his chance.

It was often near Slade where he would veer our long gray Chevrolet over to the guardrail, where the parkway plunges down and back up an escarpment at the edge of the eastern Kentucky coal field. My mother would sigh patiently as he summoned us from the car for some fresh air and a geology lesson.

I would have been embarrassed if my friends had been with me, but they weren't. So, like a good student, I'd take my sister's hand and tromp over to where my father was waiting. He would point out the layers in the rock, how they'd dip and twist and disappear. My sister, Emily, would nod. She was two years older, but because she had Down syndrome, I had caught up to her in height by the time I was seven or eight. She would smile up at my father with her sweet almond eyes, tracing the junctions with her finger. Meanwhile, I'd be crouching down on my knees, hunting the rock for fossils.

I was too self-conscious to admit it, but I found it fascinating that the ground held secrets, that life had been mummified into rock. The scale of it was tremendous. The formations stretched for kilometers, recording what it was like before birds flew through the air, before flowers adorned the world with color. When the seas were shallow, when Kentucky sat on the equator.

When I first saw ALH84001, I couldn't help but think of the crinoids preserved in the roadcuts back home. For the longest time I had thought those crinoids were crawling creatures, for they looked just like the ringed earthworms that would splay themselves across our sidewalk after a heavy rain. And I was accustomed to seeing sinuous shapes in the soft siltstones, blending in with the texture and color of the surrounding rock. Those impressions were trails in the mud left by worms, but they weren't left by crinoids. Crinoids didn't have serpentine, soft bodies, and crinoids didn't crawl. It would have been awfully hard to convince me, standing in front of that rock wall, but the slender stacks of ossicles I kept seeing were the calcified rungs of the arms of ancient sea lilies, the strange and beautiful cousins of starfish.

Most things disappear, I remember my father explaining. Hard, solid things, he said—those are what remain. Everything else breaks down and washes away.

BEFORE MCKAY HAD departed for the ancient seabed of the Texas Hill Country, he'd tucked a SkyPager into his bag, just in case anything came up. After a couple of days of tubing, he realized the pager had been completely silent, and that silence was starting to make him nervous. On August 6, he decided to call the office on a pay phone outside the camp store, just to check in. To McKay's dismay, the story of his rock was everywhere. The pager, it turned out, didn't work beyond Houston's city limits, and NASA was in a complete frenzy. Panicking, McKay's wife and three daughters rushed him to San Antonio Airport so he could fly to Washington. The very next day, he found himself under the bright lights of an auditorium at NASA headquarters, in front of hundreds of reporters.

Just as the two-and-a-half-hour NASA press conference began, President Clinton stepped to a podium on the South Lawn of the White House to proclaim Dave McKay's findings to the world: "Today, rock 84001 speaks to us across all those billions of years and millions of miles. It speaks of the possibility of life. If this discovery is confirmed, it will surely be one of the most stunning insights into our universe that science has ever uncovered. Its implications are as far-reaching and awe-inspiring as can be imagined . . ."

As McKay was introduced at NASA Headquarters, he looked befuddled in his striped shirt and space-themed tie, overmatched by the attention he was receiving. A portion of ALH84001 rested on the swath of black velvet in front of the dais. It was only 1.3 ounces, a small fraction of the four-pound rock that was collected twelve years earlier in the Allan Hills of Antarctica. When prompted, he nudged the case forward, nervously tucking his chin, and stared as the glass reflected the dazzling flashbulbs of dozens of cameras.

The directors of the National Science Foundation, the National Institutes of Health, and National Academies of Sciences, Engineering, and Medicine all sat in the front row. McKay listened as the NASA administrator boomed that it was an "unbelievable day" and as he repeated the president's call for a sweeping White House summit on space exploration later that fall. "We're now at the doorstep of the heavens. What a time to be alive!"

In the hours that followed, ALH84001 made headlines around the globe. Within days, almost a million people had seen the paper on *Science*'s fledgling website and Web alert service. Prices for meteorites soared at auction houses, leaping from two hundred to two thousand dollars a gram. Congress scheduled hearings. News crews swarmed like bees through the corridors of Johnson Space Center. After centuries of observation, after launching scores of probes, there was an answer to the question of life on Mars, and it had simply fallen from the sky.

ALH84001 HAD LANDED in Antarctica thirteen thousand years ago, but it was the handful of years before its discovery that had made all the difference. Had ALH84001 been spotted and analyzed during the *Viking* mission, scientists would have lacked any reason to believe that a meteorite could have made the trip without being shocked and heated beyond hope for any meaningful analysis or that microbes could survive in truly punishing environments. Nor would they have had the sophisticated instrumentation available to McKay in the 1990s.

By that time, we had made giant leaps forward in understanding the limits of life. However, no researcher had ever found a microbe as small as the fossil in ALH84001. McKay had a reputation for being an extremely careful scientist. His description of the meteorite's features was never questioned, but the implications he drew soon came under fire. Chief among the criticisms, called out prominently in a National Research Council report, was that those segmented compartments might not be big enough to encapsulate the biochemistry necessary for life.

Soon a young British researcher who'd been invited to join McKay's team discovered some worrying signs that the organic material in ALH84001 might be of terrestrial origin, that it might have seeped in with Antarctic meltwater. Then, not long after, another research group at Johnson Space Center, led by McKay's own brother Gordon, demonstrated that nearly identical-looking strings of magnetite crystals could form spontaneously in the laboratory. McKay quickly rebuked Gordon, arguing that artificially pure starting materials were used in those experiments and that such purity would never be found in nature. He maintained that a biogenic interpretation made the most sense and that the case for life was "further strengthened by the presence of abundant [fossil-like structures] in other Martian meteorites." His critics countered that he had been seduced by morphology. Shapes can play tricks on the eye.

As the months went on, it was hard for McKay not to take the attacks personally. They poured in from all directions, with Gordon even jesting to reporters that his brother was getting "a little testy." McKay would work all day and then return home late at night in his old Chevy van, retreating to the house he'd built on the floodplain, its walls covered with kimonos from the time he had spent with the Geological Survey of Japan. As the stress built, McKay stuck to his guns, unaware that a year after the announcement he would find himself in a hospital undergoing quadruple bypass heart surgery.

A remark that Carl Sagan made early in his career, that "extraordinary claims require extraordinary evidence," was often repeated in the wake of the controversy. By the time Clinton's White House summit on space exploration took place, Sagan was unable to stand, much less attend. He had always been the apologist for dispiriting results on Mars, the unwavering optimist, the soothsayer—but from a treatment center in Seattle, shortly before his death, he conceded that "the evidence for life on Mars is not yet extraordinary enough." Others agreed, and the idea that ALH84001 contained fossil remains of Martian life was largely abandoned.

HOW MAGICAL IT had been for those few weeks, how completely riveting. It seemed so unlikely and at the same time perfectly in line with the breathtaking revolution taking place in the world of biology, where the pace of discovery was accelerating. The human genome was nearly mapped, as were the genomes of many simpler organisms. Strands of DNA were being sequenced all over the world. Every type of living organism was being tacked onto a phylogenic tree, with extremophiles leading the way back to the very root of life on Earth. Within a few weeks of the ALH84001 announcement, genomic data unveiled an entirely new domain of life, the archaea: primitive microbes capable of surviving in extreme conditions that were previously unknown to science. With the discovery, biology's five kingdoms (animals, plants, protists, fungi, and bacteria) dissolved, replaced by a system that recognized how prevalent and diverse simple, single-celled organisms were on Earth. At the same time, Dolly the sheep was cloned. Widespread harvests of genetically engineered corn and soybeans were arriving in supermarkets. Pharmaceutical companies were fanning out across the globe in search of rare, potentially life-saving organisms, and with samples in hand, they were discovering, patenting, and commercializing bizarre new compounds. It was a heady time. And somehow fitting that a rock capable of unlocking the mysteries of life would just show up out of nowhere.

ALH84001 was a glimpse into a future brimming with possibility. What if life on Mars was completely different from life on Earth? Every organism we knew, down at the molecular level, was just the same: DNA-based, with DNA coding for RNA, RNA coding for amino acids, amino acids coming together in proteins, and proteins building cells. What if those tiny Martian cells were built on an entirely different biochemistry? Perhaps the answers were inside a meteorite. Perhaps the rock would reveal the underlying constitutional nature of life—perhaps even evidence of a separate genesis. Or, if those cells bore fundamental similarities to life on Earth, it might suggest universal laws of biology, just as there were universal laws of physics and chemistry. In these ways, when ALH84001 fell to Earth, it had landed us on the brink of discoveries so profound that they promised to transform the very nature of science.

The possibilities didn't end there. Even if life on Mars was exactly like life on Earth—ancestrally related, caught like a cold from the next planet over—that too would be revelatory. A microbe that had hitched a ride in a meteorite could speak volumes about the nature of evolution, allowing the tape to be replayed, the adaptive and random splitting of lineages to be re-charted. It could open a window into how differently things might have turned out for us here on Earth. Maybe Martian phylogenetics would also indicate that life started in some warm little pond? Or maybe we would discover that we didn't need to find Martians, that we *were* Martians? After all, countless tons of rocks, rocks like ALH84001, were exchanged between the planets early in their histories, and more meteorites were lobbed from Mars to Earth than Earth to Mars, as debris sailing about our solar system was tugged toward the sun.

In the end, ALH84001 was not what David McKay hoped it would be. But for a moment, we held in our hands what we had been after for so long: a Rosetta stone for biology.

CHAPTER 6

Traversing

THE FIRST INTERPLANETARY rover, the size of a suitcase, came blazing through Mars's thin atmosphere in the summer of 1997. It had been two decades since we'd been to Mars, and we had entered a new age of robots. The *Pathfinder* mission was designed to test how a rover might roam across the Martian surface, remotely piloted from mission control two hundred million kilometers away. The rover was a solution to a problem as old as exploration itself: How do you study something that is very far away? Logistics become exponentially more complicated the farther you go, introducing new constraints on what can be carried, assembled, and accomplished. Lunch in a kitchen is easier, and can be more elaborate, than lunch on a mountaintop. Never was this more of a problem for science than with *Viking*. So much energy and effort had gone into simply getting to the surface of Mars, and yet after all those years of planning, the lander's instrument payload could only be deployed to investigate what happened to be right in front of it. As a result, the subject of *Viking*'s study was a matter of chance, and there was no way to know if better scientific targets were just out of reach.

Pathfinder was meant to change that. The rover would usher in a new age of agile, real-time planetary exploration, one where each of the ongoing observations would enable adjustments to the traverse, allowing scientists to capitalize immediately on the data returned.

The spacecraft was the first in a series of low-cost missions developed under NASA's late-1990s mantra of "faster, better, cheaper." It was designed under the leadership of an irascible NASA administrator from the South Bronx, a former aerospace-industry executive. He was determined to show that we could not only return to Mars but we could do so for a fifteenth the cost of one of the *Viking* missions, in half the time, using a team a third the size. In service of that goal, NASA had jettisoned the idea of slowing down in orbit for a soft landing and instead decided, for the first time, to barrel straight into a planet.

As *Pathfinder* approached Mars, it was the dead of night on the planet's surface. Because the landing site was rotated away from the sun and away from Earth, the Mars scientists who had waited twenty years since *Viking* to get back to the Red Planet would have to wait until sunrise on Mars to know if the little rover was safe. They gathered in mission control, holding their breath as the landing sequence commenced: the spacecraft lurching toward the surface, a parachute deploying at supersonic speed. If all was well, a flotilla of protective airbags would inflate as the probe slid down a twenty-meter Kevlar tether. The spacecraft would then thud onto the ground at highway speed, glancing off the surface like a giant beach ball.

Hours later, when at last the data began to trickle in, the team realized that *Pathfinder* had bounced over fifteen meters into the air, bounced several more times, then settled itself on the surface of a new world. The crash landing had worked—the airbags had miraculously cocooned the rover.

But *Pathfinder* wasn't out of the woods yet. The mission wouldn't be a technological success until the rover powered up, drove down off the lander, and rolled across the surface of Mars. And so it was to the team's great dismay when the first, dramatic images came back to Earth and showed one of the airbags still billowing in the air. That was a problem: The pyrotechnic devices in the latches of the lander panels were meant to fire and pop free after the spacecraft bounced to a halt, allowing the lander to peel open like a flower and release the vehicle that would actually explore the surface, the *Sojourner* rover. Winches should have already pulled the airbags back in, clearing a path for the rover to drive down a ramp onto the red ground. Instead, the errant rippling airbag was blocking the rover's descent.

In the hours that followed, mission control programmed one of the petals to lift up and flap down again, hoping to clear the finicky airbag from the rover's path. But during the night, a far more crippling problem arose: Communications between the station and its small companion began breaking up. *Sojourner* carried nearly the entire scientific payload; the landing station itself was lightly equipped—it had only a camera, three wind socks, and a radio. But the landing station was the rover's sole link to Earth. On its own, *Sojourner*'s voice could reach just a few hundred meters.

The engineers worked the problem, sending commands to switch the radio electronics on and off again periodically for the next twelve hours. Though they never pinpointed the cause of the glitch, the engineers were eventually able to reestablish the radio link, to the point where about 80 percent of the information was coming through. They also cleared the finicky airbag. Even though its first steps were the hardest, *Sojourner* finally stood up and trundled off the ramp. The rover's six small wheels, suspended on rocker-bogies, inched into the craggy swales of Ares Vallis.

Cameras clicking, the rover captured images of boulder-strewn ridges and jagged flood debris, rocks everywhere. The team knew the mission had touched down in a giant outflow channel. The rocks were thought to be from far-off places: from the hummocky hills of Margaritifer Terra, from the tangle of Iani Chaos, from the highlands of Xanthe Terra. They were all out of place, but they might each tell a story.

Within days, a measurement on "Barnacle Bill"—the first chemical analysis ever made of a rock on Mars—was already starting to reveal a shockingly turbulent past. Barnacle Bill appeared to have been formed from relentless cycles of melting, solidifying, and remelting, meaning that Mars was once characterized by immense heating and internal stress. As the summer progressed, *Sojourner* trundled on to other nearby rocks—including "Yogi" and "Scooby Doo"—which the mission named after cartoon characters.

Surprisingly, the pictures coming back showed rounded pebbles strewn about the Martian terrain, and rounded sockets in the rocks too: evidence that they once tumbled through running water. As *Sojourner* explored, it also discovered sand piled into fluted patterns and, in the distance, cresting into dunes. That meant wind, not just water, had played a powerful role in creating and shaping the planet's enormous landforms—in forging its bulwarks, molding its attributes. As the rover explored, the scientists working the mission started to realize that they were dealing not just with the landscape spread out before them but also with a highly dynamic history. Apparently, the forces on Mars were once strong enough to tumble the edges from rocks. To pick up the smallest pieces of its world and move them across impossibly large distances.

I WATCHED THE *Pathfinder* mission unfold from a span of old-growth cane forest. I was seventeen years old that summer, and I too was barreling headlong into a new existence. I was going forth, off to begin adulthood, lifted like a piece of spall. I'd spent my entire life in Kentucky, barely ever crossing the state line, but soon I'd start college. I'd already left my hometown: the sweetgums in my backyard, the little brick house where I'd always lived. I'd packed my things into milk crates and driven the hour and a half to Camp Piomingo, where I was spending the summer working as a counselor along the sun-dappled banks of Otter Creek. I had a bunk in an old wooden cabin, and each night, the woods pressed in close. I'd fall asleep to the sounds of katydids and cricket frogs. I was assigned to watch over the counselors-in-training, even though I was new at this myself and just barely older than my charges. I spent my days leading them through caves and on creek walks, shivering as water snakes swept past our limbs. We explored ridges and gullies, repaired trails, and rested in the shade of the rocks down in Doe Valley.

The day *Pathfinder* landed was the Fourth of July. I was celebrating our country's birth, and my own newfound independence, by lighting sparklers at a party on the cabin porch. While I could find Mars in the sky, a steady red drop of light, there hadn't been a Mars mission in my lifetime. I'd read stories about the rock from Mars the year before, and I'd heard about NASA's plans for a new wave of robotic missions. It'd been thirty-two summers since Mars exploration had begun, thirty-two summers since my father had read about *Mariner 4* on the *Courier-Journal*'s front page. Now the *Courier-Journal* headline declared that we were BACK ON MARS, in letters almost as large as the masthead itself.

In the days that followed, I grabbed the newspaper in morning mess hall. I read about the itinerant rover, named for Sojourner Truth, the former slave who escaped captivity and went on to be a prominent abolitionist, and the rough, rocky floodplain the rover would explore. I read about the engineering systems, about how *Sojourner* would periodically stop driving and send a "heartbeat" message back to the landing station.

One of the articles closed with a quote from a professor at Washington University in St. Louis, a *Pathfinder* scientist working on the very same campus where I would soon be studying. I underlined his name—Ray Arvidson—and carefully tore the article from the page to send to my father, who was certainly following these distant goings-on. I hoped the fact that I would be at a university with a famous planetary scientist would make my parents feel better about the debt they were about to incur to cover the daunting gap between my scholarship and the full cost of attending.

Once on campus, I made a beeline for Ray's classroom—that fall, he was teaching a course called Land Dynamics and the Environment. From there it was a short hop to his lab, where he did his pioneering Mars work. Ray was nearly fifty, a Swede who'd grown up in New Jersey. He was humble, soft-spoken, and tremendously respected, having just taken over as the department head of Earth and planetary sciences. He had a short beard that was starting to turn gray, and his eyes crinkled whenever he smiled. He specialized in remote sensing: how to get beyond your immediate environment, how to discern the nature of a place from a distance.

On his flickering desktop, he showed a group of us how to "see" beyond the visual wavelengths, into the ultraviolet and infrared. Computer techniques stretched and transformed orbital views into a psychedelic swath of colors. He would toggle a few settings on his display, and types of land cover that initially all looked the same would filter in and out, revealing their underlying complexity. We were astounded that by simply adjusting the wavelengths of the light he looked for, he could pinpoint dozens of rocks and minerals that had previously been hidden.

Ray had grown up shooting off balloon rockets in his backyard. As a graduate student at Brown University, he'd analyzed the *Mariner 9* data, then worked on the *Viking* landers, taking over as the head of the imaging team after the first year of operations. I particularly loved listening to his stories about the *Pathfinder* mission.

In one of those conversations, he explained how the simple fact that we'd finally returned to Mars had allowed the team to calculate a crucial variable: the planet's moment of inertia, which scientists considered the "single most important number about Mars that we didn't know." By triangulating *Pathfinder*'s position with the location of *Viking*'s landing twenty years earlier, the team could determine the degree to which the planet's spin axis had changed, like the wobble of a spinning top—which helped them to calculate the moment of inertia. With that, the team could discern how mass is distributed in the center of the planet, revealing how Mars formed and how it changed over time. With one landing and a little math, he said, we could peer into the planet's interior.

What we learned, Ray explained, was that Mars had to have a dense metallic core. Previously, no one knew whether the planet would have been warm enough to differentiate into layers, but this result indicated high heat flow in the past, which would have warmed the surface, triggering active volcanism. Volcanism, in turn, would have spewed out greenhouse gases, thickening the atmosphere. Higher heat flow also meant that it was conceivable that Mars once had a molten core with an active core dynamo, which would have led to a magnetic field that protected the surface from harmful radiation. And suddenly, Ray said, we understood that Mars might once have been a place with a warm surface, a thicker atmosphere, *and* a protective magnetic field—the kind of place where life might have taken hold. Like most of my classmates, I neither completely followed nor fully appreciated the science. We were, after all, first-semester freshmen, doing well to find the dining hall. But it did make an impression on me that we could somehow know so much about a place from just knowing where we were—and where we had been.

WHEN RAY INVITED me to stay on after my freshman year to work in his research lab with a small stipend from the Missouri Space Grant Consortium, I couldn't believe my luck. Ray was working with NASA on a new method of mobile exploration. With colleagues from JPL, he was developing a prototype of a planetary payload that could map at far lower altitudes than an orbiter and in the meantime collect atmospheric measurements. This "aerobot" was designed to be piloted through the skies of another planet or moon—Mars, Venus, Titan—like a tiny mechanical hot-air balloon. The idea was to close the gap between the observer and the observed.

To test the aerobot in the sky, Ray had teamed up with a university alum, an adventurer named Steve Fossett, who was attempting to become the first person to circumnavigate the globe solo in a balloon. It was aviation's "last great challenge," and surely its hardest, leaving the pilot completely at the mercy of the winds. He'd chosen to fly in the southern hemisphere, despite the risks of such desolate stretches of ocean, to avoid the political difficulties of crossing China, Iraq, and Libya. Fossett was a fifty-four-year-old commodities broker who'd made his fortune on the Chicago exchange. After he'd tired of finance, he swam the English Channel and climbed most of the world's tallest mountains. He'd completed the Iditarod dog race. He'd set dozens of records for speed sailing and had even flown a glider into the stratosphere.

To some, it seemed unusual that Fossett's next great adventure would involve Mars science. But he was by no means the first person to experiment with balloons as a way of advancing planetary exploration. Preceding him were several lionhearted Mars explorers, including David Peck Todd, the leader of Lowell's Chilean expedition. Not long after he returned, Todd announced to *The New York Times* that he planned to ascend in a balloon to the highest possible altitude a human could reach, wireless device in hand, in an attempt to communicate with Mars. The idea of "hearing" Mars had been very much in vogue. Scientists as prominent as Nikola Tesla and Guglielmo Marconi, the rival inventors of radio communications, had both been fascinated by Mars. Having read Lowell's reports of intelligent life, they had tinkered with methods for detecting Martian radio signals. Tesla, captivated by the potential to use radio to communicate over unthinkably long distances, once remarked that Mars was only five minutes away by wireless. Todd, however, went further: He enlisted the help of Leo Stevens, a famous aeronaut, to prepare a metal box "made of aluminum for lightness" that he could be shut into. It was fitted with a machine to drive out carbonic acid gas and supply oxygen with air pressure. Todd's hope was to rise above the crowded airwaves, thereby giving himself the best possible chance of communicating with our near neighbors.

Even though the Aero Club of New England offered to let Todd use an enormous balloon called the *Massachusetts* for his Mars expedition, his grand plans never came to fruition. He was slowly eased off the Amherst astronomy faculty after the demise of Lowell's canal hypothesis, which he had championed. But he still clung to the ambition that radio might bring a Martian civilization within our reach, even convincing the U.S. Army and Navy to shut down all radio communications for two days in 1924 in what became known as "the Big Listen."

Nearly thirty years later, another balloonist proposed a still more ambitious experiment. Audouin Dollfus was one of the only practicing planetary scientists in the early 1950s. Mars had been all but ignored for decades, but Dollfus was desperate to know if there would be enough moisture in the Martian atmosphere to support simple forms of life. He had perfected the use of prisms to separate the light of faraway worlds, to measure attributes like water vapor from absorptions in the infrared wavelengths, but Earth's own atmosphere interfered with his attempts. Like Todd, he was trying to understand

Mars in a new way, using waves beyond those our eye could see. And like Todd, he was wrestling with the problem of being stuck on the surface. To answer his question, he would have to find a way to rise above Earth's damp air.

Dollfus was a Frenchman—an adventurer at heart, small and slight, with bright eyes and pink cheeks. In 1954, he and his father completed the first astronomical observations from a balloon, but he couldn't get a measurement of the Martian atmosphere. He concluded that he needed to get twice as high—into the stratosphere. He began building an airtight gondola attached to a telescope with a foot-wide mirror. In 1959, he suspended both from a cascade of over one hundred weather balloons, clustered along a few hundred meters of nylon cable. He insulated the metal sphere of his gondola with foam rubber to protect him from the bitter cold.

He transported the capsule, slung beneath a French Air Force helicopter, to a military airfield. A flock of assistants inflated his white balloons, filling a field with billowing bags of hydrogen, any one of which could have caught fire. Carefully, they were tied to the long nylon cable in groups of three, extending five hundred meters in length. Dollfus climbed through the manhole, and when the last of the bunch was secure and sufficiently far away, a small explosive charge severed the anchoring cable and sent the cabin into the air. Aircraft approaching Paris were warned not to fret about the strange hazard—what looked like a string of Spanish onions.

Dollfus launched right around sunset. As he floated up into the stratosphere, he saw a perfectly horizontal line dividing the sky. He wrote in his logbook that the air below glimmered with dust, resembling an almost phosphorescent sea, whereas the sky above was perfectly pure and dark despite the full moon, the constellations shining without scintillation.

Dollfus spent part of the night at an altitude of fourteen thousand meters—forty-six thousand feet—before a gust of wind caused a group of balloons to burst and the cable to shear. He descended steadily through the darkness, thudding down in a cow pasture near the village of Nivernais. In the end, he still failed to get his measurement, but what a brave attempt he'd made! By reaching those extraordinary heights—decades before the invention of space telescopes like the Hubble—Dollfus cracked the door to studying astronomy from space.

AT EIGHTEEN YEARS old, I was completely taken with the idea of the grand gesture. To me, Fossett's *Solo Spirit* mission seemed like the next daring adventure, with the potential to change the way we might go about exploring Mars's atmosphere. Engineers had carefully affixed our aerobot payload to his gondola, which was shipped to Argentina in the run-up to launch. Back in St. Louis, mission control was set up in a tall Gothic building on campus, in what felt like the top of a castle. There were wooden doors and wood-paneled walls, tables with desktop computers, banks of phones, clocks with different time zones, navigation books, and tables of data. The aerobot had sensors for position, temperature, atmospheric pressure, humidity, vertical wind velocity. All the raw data from the payload telemetry—everything the white box, gray fan, and wired sensors detected—would be sent back to St. Louis through a satellite.

When Fossett's balloon lifted up from a soccer stadium in Mendoza, I cheered wildly in mission control. I could only imagine the sense of excitement and invincibility Fossett must have felt as he soared into the sky. He had all the latest technologies—GPS devices, a fax machine, and a satellite phone. He rode in a futuristic capsule, fashioned from a Kevlar-reinforced carbon-fiber composite. His balloon was a special Rozière design, which utilized novel temperature-control features to keep his helium gas from cooling on chilly nights. He'd attempted this circumnavigation before, but this time felt like his lucky break. I was certain that both Fossett and our little aerobot would succeed marvelously.

The balloon made swift progress from Argentina out over the Atlantic, arriving in Africa in just a few days, then gliding effortlessly over the Indian Ocean. I traced the balloon's route on a giant Mercator map, pressing small red pushpins onto the coordinates that the aerobot beacon signaled, as thousands of kilometers away, Fossett gazed upon beautiful vistas. I imagined him transcendently aloft, on the cusp of overthrowing the sky's "last great challenge," of finally winning the record.

The balloon was making fantastic time, and the aerobot was providing a steady stream of information. Geopositional data were relayed every ten seconds, atmospheric measurements every minute. The positioning of an antenna within the center of the balloon had worked extremely well in terms of boosting the signal. Observations matched perfectly with what was observed by satellite data. Ray's prototype would hopefully experience similar success on Mars one day.

Back in St. Louis, our shifts ran for eight hours, twenty-four hours a day. One of my jobs was to check the aerobot data as it was relayed to mission control, and I was among the first to notice when the signals stopped. Initially I thought it was a technical issue, but somewhere between Queensland and New Caledonia, in the early-morning darkness, a thunderstorm had filled the sky. Fossett's balloon had begun a rapid ascent, and as it rose, it ruptured.

The balloon fell through a sky spangled with lightning—careening almost nine kilometers to the ocean below. Sheets of hail barreled into Fossett as wind tore at the fissure in his balloon. He frantically pushed tanks of fuel overboard, desperate to slow his descent. His capsule still slammed into the sea. It was quickly pulled down under the surface, filling with water. Almost simultaneously, the propane burners burst and caught fire. Fossett fought his way out, but he was left bobbing, all alone in the shark-infested waters of the Coral Sea.

I'd heard his wife's voice for a split second, wavering, when I'd answered the phone that morning. I quickly passed the receiver to one of the navigators who was advising Fossett from mission control. I stood back, my eyes wide, as he explained what we did, and didn't, know. Since I'd worked the night shift, I hadn't slept, and I didn't sleep all day. I sat in the back of the room as the navigators searched for a locator beacon, as the reporters began to call, as the public began to gather the facts. It was dizzying.

Fossett almost died out in those uncharted reefs. He activated that beacon twice, but then the satellite stopped detecting the signal for a few hours. Cold and terrified, he was finally spotted by a French plane, then ten hours after that, he was pulled onto the deck of a boat by an Australian yachtsman.

Fossett's entire journey had been more precarious than I'd realized. He'd been relying on oxygen tanks and barely sleeping. At one point, he singed his eyebrows and then ran out of toilet paper. It was a grueling experience for him, and for his family down below. From the safety of a rescue boat, he told a reporter he might just "sit back and smell the roses for a while" before flying again.

FOSSETT WAS ON my mind when I flew for the first time over an ocean, as a sophomore in college. As I stared down at the vast and endless sea, I couldn't believe how far it stretched—the unbroken expanse, the sheer emptiness of those waters. I felt a deep sense of relief as the Big Island of Hawaii came into view, its cliffs rising like a prehistoric pelican, like something out of *Jurassic Park.* Finally, a foothold.

I was journeying to a volcano as part of a class trip led by Ray. It was the second geoscience course I'd taken with him, and a tremendously exciting one. Like most of my classmates, I had never been so far from home. After we landed in Hilo, the world turned from sea to rock. We headed in the direction of Kilauea, in rented vans. I pressed my face against the window as we drove the Chain of Craters Road, as we wrapped around pahoehoe lava and 'A'ā flows, right down to where the molten rock flowed into the sea. The sky turned to fire as the sun set, then it was the blackest night I'd ever seen. Constellations, which had always seemed implausible to me, suddenly made sense, their missing stars bright against the enormous sky.

We left Kilauea a couple of days later and made our way to the desolate summit of Mauna Kea, a dormant shield volcano on the other side of the island. We stopped two-thirds of the way up to acclimatize for a few hours before continuing to the summit, a disorienting forty-two hundred meters in altitude—fourteen thousand feet—where the air carried 40 percent less oxygen. As the road climbed, we passed the tree line, then the last of the scrub and the last of the lichens, until we were above even the clouds. The landscape was gray and red and black in every direction; in places it even smoldered with a sheen of purple. There were shards and ash and cinder cones. It felt like a bruise, crystallized in the world.

One day, when everyone was having lunch, I wandered over to check out the view from a distant ridge, where the solid lava gave way to pyroclasts and tephra. Without really noticing, I was kicking at the rocks as I stepped. I overturned a surprisingly large one with the toe of my boot, and as my eyes fell to my feet, I startled. Beneath the vaulted side of that adamantine black rock, a tiny fern grew, its defiant green tendrils trembling in the air.

There in the midst of all that shattered silence was a tiny splash of life. I crouched down to see it better. Here was a piece of the world I'd left behind. Here was a piece of my childhood. Every summer on Pine Mountain, about thirty kilometers south of Hazard, my sister and I would race down a side spur onto a footpath and into thickets of mountain laurel and rhododendron. We'd scramble down the living stairway of a century-old tulip poplar that had been felled into a gorge but hadn't died, holding upside down to its roots, even after hand-hewn steps were carved into its immense trunk. Those beautiful, breathing steps would deliver us into a hemlock ravine, and from there we'd follow the roar of the streams into the riotous fractal heart of the fern garden. We'd find a mossy rock and sit our small bodies down, resting among the fronds until our breathless parents caught up. There were ferns taller than us and ferns tinier than our fingernails, all with intricate patterns, spreading their palms into the greenest symphony I'd ever seen.

That fern on the volcano was even more striking up there by itself, all alone. It was just so impossibly triumphant. I couldn't pull myself away; I looked at it for so long that the others had to come find me. I showed it to them, but I didn't have the words to explain its beauty, its significance. I couldn't tell them that somehow, huddled under a rock, growing against the odds, that fern stood for all of us.

Even though I couldn't articulate it, I had an inkling then of what I know for certain now: There was something in that moment that made me become a planetary scientist. It was then, on that trip, that the idea of looking for life in the universe began to make sense to me. I suddenly saw something I might haunt the stratosphere for, something for which I'd fall into the sea. Not fame or glory, or a sense of adventure, but a chance to discover the smallest breath in the deepest night and, in so doing, vanquish the void that lurked between human existence and all else in the cosmos. On that trip, I started to realize that, just like with *Pathfinder,* the process of reaching might tell me the most, might give me the chance to grasp the deepest mystery. In finding that fern, I also found something small, fragile, and worth cultivating deep inside myself.

When I got back to St. Louis from the Big Island, I tossed a hunk of the volcano into the hand of my best friend. It was part of a gathering of rocks on her desk. It was a motley collection—a pumice here, a sandstone there—but it told her that I was finding my direction. As for me, I couldn't help but think the tableau looked a little like Ares Vallis strewn with bits of the known world. I was beginning to feel, for the first time, like a real explorer.

CHAPTER 7

Periapsis

I WAS A SOPHOMORE in college, one of thousands attending a crowded scientific conference, when I first saw the kaleidoscopic new map of Mars made from data collected by the *Mars Global Surveyor* mission. I'd taken my seat alongside some other undergraduates in the packed ballroom as an MIT professor named Maria Zuber walked to the front of the room. She seemed impossibly small standing behind the podium. Then her slides blazed to life and she began to speak.

I remember feeling a little distracted at the beginning of her talk, as her voice filled the cavernous hall, and I puzzled over it for a few seconds. Then my back straightened as it became clear what I was responding to: It was the first time I'd ever heard a woman give a planetary science talk.

As I willed myself to focus, I realized that Maria was delivering a spectacular explanation of the mission science—better than any I'd ever heard. Her confidence and enthusiasm radiated, and the audience was locked in. Sitting there, suddenly aware of our shared status as women in a roomful of men, I couldn't help feeling a swell of pride, as if she was somehow speaking for me and the other aspiring female scientists in the room.

Then she clicked to the magnificent map. She paused for a moment, letting it linger on the screen behind her, as if she knew the effect it would have. She gave us all a moment to soak it in, to reckon with how different it was from anything that had come before. The usual cinnamon surface of Mars had been swallowed by a rainbow of color, magnificently delineating topographic contours. The iridescent surface was cut with canyons and chasms and spiral troughs. The northern hemisphere was as smooth as the abyssal plains of Earth's own seabed—tantalizing evidence of an ancient ocean. There were continuous layers of bright and dark bands alternating right down to the edge of the ice cap, a record not only of changes in the seasons but also of long-term climate patterns. None of us had ever seen Mars like this—a place we could almost touch. Her rendering had thrown the planet into exquisite relief, flinging two dimensions into three. Meridians arced down from the pole, like strips of lead across the colorful sphere. In that darkened room, it shone to me like a church window.

THAT MAP OF Mars was a long time coming, both for NASA and for Maria. The instrument that collected the data was named MOLA. It stood for the "Mars Orbiter Laser Altimeter" and also, as was sometimes pointed out, the type genus of a family of "strange, large oceanic fishes." MOLA was originally proposed in the 1980s as part of a NASA mission called *Mars Observer.* That spacecraft was meant to be the first in a line of "planetary observers," orbiters based on commercial communications satellites that NASA could buy on a fixed contract. The plan was to use the space shuttle, which was already ferrying astronauts to low Earth orbit, to launch the orbiter.

The *Mars Observer* mission was a big break for Maria, fresh off her PhD, her first chance to work on a spacecraft heading to Mars. She had grown up amidst the coalfields of Carbon County, Pennsylvania, a place where prosperity and economic opportunity had dried up with the decline of coal mining. No one in her family had gone to college, and her parents had difficulty understanding why anyone would want to stay in school as long as she did. But she'd been fascinated by space from the time she was small. She'd jump up and down in her playpen, pointing to rocket launches on TV. She even loved the scenes of mission control.

As she got older, she started watching copious amounts of *Star Trek*—she gravitated to Lieutenant Uhura—and spending as much time as she could alongside one of her grandfathers in his old garage. He'd left school after eighth grade and had suffered from black lung disease and underemployment most of his life. Yet at one point, he'd scrimped and saved his meager earnings to buy a telescope. She didn't know that at the time, for he no longer had it. But he had learned to make his own, and he taught his granddaughter the craft. By the time she was ten, she'd learned how to grind mirrors; she would set them up in her backyard and stand out there for hours, often shivering in the cold, inspecting the night sky.

There wasn't much money for college in the family of a state trooper with five children. As Maria neared the end of high school, her guidance counselor encouraged her to apply for a scholarship to Penn State. After selecting her, the scholarship board called her guidance counselor to instead suggest the University of Pennsylvania. Maria matriculated that fall, helping make ends meet by operating telescopes for visitors at the Franklin Institute in Philadelphia. In 1986, when she graduated with a PhD from Brown, she was the first person from her high school to ever earn a doctorate.

As a graduate student, Maria worked on theoretical models of planetary evolution. The data set she needed to test those models wasn't available, so she decided to collect it herself. As a newly minted researcher at NASA's Goddard Space Flight Center, she worked with the team for one of the instruments on *Mars Observer,* a radar altimeter to measure the physical landscape—the depths and heights of the Martian terrain. But after the *Challenger* shuttle exploded with seven astronauts aboard, NASA put the mission on hold. *Mars Observer*'s launch was delayed by a couple of years, and soon instruments were being jettisoned to control the ballooning costs. NASA still wanted an altimeter to fly, but it could no longer afford the one being built. The agency announced it would hold an open competition for a cheaper version.

Maria knew that, as part of Star Wars—Ronald Reagan's Cold War missile defense system—the United States had been investing billions of dollars in laser technology. She finagled a security clearance, and she and some other young scientists started meeting with engineers. They had already sorted out the power-supply issues, nailed the pointing, and figured out a way to stabilize the jitter. All the planets were still being mapped with radar, but Maria realized that a laser system, if she could get it to work, would blow that method out of the water.

The idea was to measure the distance from an orbiting spacecraft to the surface of Mars and use that to calculate the altitude of the Martian terrain. The instrument would shoot laser bursts, each lasting only eight billionths of a second, and then very accurately time their reflections. Even though the actual power in the laser was just a fiftieth of that in a refrigerator light, the instrument was designed to render arrestingly accurate readings of topography.

When NASA eventually approved the $10-million proposal, it was taking a huge risk: It wasn't entirely clear that this idea would work. The pointing of the gold-coated beryllium telescope had to be extremely precise, as did the timing, down to nanoseconds. The instrument had to have its own temperature-controlled clock, cooled to avoid drift. Even the slightest error would be difficult to correct. There were other obstacles too. When the engineers wouldn't let scientists into the cleanroom, Maria had to get certified in laser safety as well as laser engineering. She worked on the mission for years—while she got married and had two children, whom she would sometimes need to strap into carriers and take to meetings. When time came for the launch, she brought along her toddler and infant son, who, upon hearing the rocket rumble, drowsily opened his eyes, looked around, then fell quietly back to sleep.

When *Mars Observer* at last began snapping its first far-off pictures of Mars, about a month before its arrival, Maria could almost taste how good the data would be. Precision was her strong suit, and she was certain that she and her colleagues had succeeded in building one of the world's most exact scientific instruments.

THREE DAYS BEFORE *Mars Observer* was scheduled to enter its orbit, in late August of 1993, Maria was out buying groceries. She and the leader of the MOLA instrument team were about to leave for JPL, but she wanted to stock up on food and drinks so the rest of the team could join the celebration remotely. As she was returning from the store, she got a call saying that the orbiter had gone silent. A short communications outage had been planned right before atmospheric entry, so she wasn't all that concerned. It seemed the engineers were just having some trouble reestablishing the link.

Boarding her flight to California, Maria was hopeful that rebooting the computers would do the trick—that the problem would be solved by the time she landed in L.A. But when she reached JPL, the situation was unchanged. As the hours passed, she became more worried. New commands were being sent every twenty minutes, with the deep-space tracking antennas in Canberra, Madrid, and California all focused exclusively on trying to reestablish contact. Maria and her colleagues clung to the hope that the automatic controls would still perform, firing the rockets to enter into orbit. If the spacecraft didn't slip right by the planet, it would buy the engineers more time to work on the problem. Infrared military and science telescopes in Hawaii hastily organized to try to detect the heat of the distant maneuver, but when the moment came, clouds in the Pacific obscured the view. Time continued to tick by, with each hour feeling more and more desperate. A week later, there was still no news. The spacecraft had vanished like a phantom.

Maria finally flew back to Maryland. She ended up with the last seat on the last flight of the day, a middle in the very back row. She had been three days away from getting an extraordinary stream of data and now she had nothing. When the flight attendant saw the devastated look on her face and asked what was wrong, Maria confided that she'd lost her spacecraft. "I hope you find it," the flight attendant said when Maria disembarked, slipping her a full bottle of wine.

The next day, when Maria went into her office, she spent a whole afternoon just staring out the window. She realized she had nothing to do. There was no black box to recover, no telemetry data to help understand where the spacecraft had veered off course. An investigative panel assembled by the U.S. Naval Research Laboratory couldn't replicate a single failure mode as they tested piece after piece of the spacecraft design. The following January, long after Maria had resigned herself to the fact that the *Mars Observer* would never be heard from again, the panel announced that a rupture in the propulsion system as it powered back up after the long cruise was probably to blame, a piece of tubing tucked down under the spacecraft's thermal blanket. The subsequent roar of fluids had most likely forced the spacecraft into a catastrophic spin.

IN THE MONTHS after the heartbreaking loss, Maria and a Cornell scientist named Steve Squyres teamed up to try to get funding put back in the budget to refly the mission. They started spending a lot of time on Capitol Hill. They would make appointments and show up in suits, pleading with members of Congress. Yes, the government had spent money on hardware, but it had mostly spent money on people and figuring out how to do things no one knew how to do. Now they just needed new hardware. And in a move that would have pleased William Pickering, who from his mountaintop observatory in Jamaica had penned those regular missives about Mars seventy-five years earlier, she'd tell skeptics to go home and ask their kids what they would think of having regular weather reports from Mars. That kind of inspiration, she told them, was the promise of this mission if it could just fly.

Eventually it started to look like Maria and Steve would have another shot. NASA was developing a new program for exploring Mars, based on smaller, less pricey spacecraft. A new mission called *Mars Global Surveyor* was selected for the 1996 launch window. Like *Pathfinder,* which would also launch in 1996, *Mars Global Surveyor* was part of

NASA's new "faster, better, cheaper" program. It was lighter than *Mars Observer,* only half the mass and size, but it would carry five of the same instruments, including MOLA. Maria would be the second-in-command of the instrument.

Just before launch, a reporter called Maria and asked how it felt to be the only woman among the eighty-seven investigators on the mission's science team. The question took her aback. How could that be true? But as she quickly scrolled through the names on the team roster, it became clear that the reporter was right. She realized that she must have stopped noticing things like that a long time ago. She had trained her thoughts entirely on bigger problems, like how to map Mars with breathtaking resolution—how to transform planetary cartography.

WHEN *MARS GLOBAL Surveyor* lifted into the sky in November 1996, Maria was in Florida. Her two children couldn't have been happier because, to them, a launch meant a trip to Disney World. Everything seemed to be going according to plan. About an hour after blastoff, however—up in the silence, in the great vacuum of space—a tiny lever sheared off one of the spacecraft's winged solar arrays as it tried to open. There was no sound as it slowly clinked along. It was no larger than a carabiner, but by some coincidence of physics, the trajectory sent it right into the five-centimeter hinge between the shoulder joint and the edge of the solar panel. With a tiny lever stuck in its hinge, the solar panel protruded away from the sun at an awkward angle, a good twenty degrees short of its fully open position.

There was still enough solar power to reach Mars; the complications would come after arrival. The spacecraft had been designed to enter orbit using a radical new technique called "aerobraking." The idea was to use the drag of Mars's thin atmosphere against the solar panels to slow it down naturally, rather than drawing on precious fuel. This had shrunk the cost of the rocket by a factor of five, but it left little margin for error. And with a contorted solar panel, it wasn't clear whether the aerobraking would work.

Several times during the 309-day cruise, the engineers tried to wiggle the panel back and forth to free the stray piece of metal, but they had no luck. When the spacecraft finally reached Mars, the team decided to gently "walk in" to the atmosphere to see what would happen. By this time Maria had moved to Massachusetts. She'd been offered a senior tenured-professor position at MIT—something she couldn't refuse. The university was putting an enormous amount of faith in her. The chair told her, "After *Mars Observer* failed, you should have disappeared. But you didn't, and we think that's for a reason." She had barely moved into her new office when she got a phone call from mission control. The voice on speakerphone sounded concerned. "Well, the solar panel may have broken off. We'll have to wait for the telemetry." After Maria hung up, her secretary blanched and walked out of the room, avoiding eye contact. Maria took a deep breath.

The panel hadn't in fact snapped, but it easily could have. It had started to jolt, first stopping at twenty degrees, then flapping beyond its fully extended position, like a hyperextended knee. Terrified by the weakness in the joint, the team immediately lifted the spacecraft back above the atmosphere to figure out what to do. Tests soon showed that the springs holding the panel open wouldn't withstand the atmospheric forces for long. It left the team with a big problem: *Mars Global Surveyor* needed to slow down dramatically, trimming its huge egg-shaped ellipse into a tight circular track. The only way to get into a mapping orbit was to aerobrake, but they couldn't aerobrake without risking the solar panel.

The engineers knew that fully closing the solar array wasn't an option, as it would generate too much heat right up against the spacecraft, so they decided to tilt the panel so that it was in line with the atmospheric drag, thereby reducing the air resistance by two-thirds. As they sent the command to perform the tilt, they were aware that it would come with an enormous cost: It would mean looping an extra few hundred times around the planet before reaching a circular orbit, gently grazing the atmosphere each time the spacecraft swooped in close to the surface. The primary mission couldn't begin until aerobraking was complete and Mars drifted back into the correct alignment with the sun. This would mean waiting another year and a half.

MARIA HADN'T SET out to be a mapper, but she'd always admired maps of Mars, including those that were drawn long before she was born. Their resolution was coarse by her standards, unable to capture the kind of details that would bring the planet to life, but the names of the features were full of meaning. Isidis, Arcadia, and Elysium—they all evoked a mythic, ethereal place.

Most of the nomenclature of Mars had been developed more than a hundred years earlier, by the same nineteenth-century Italian astronomer who had identified the swarm of crisscrossing lines on the planet's surface. Giovanni Schiaparelli also saw scores of other features that no one had seen before, features in need of names. Like Maria, he had been peering into space since he was a small child. He'd also grown up in a large family in a quiet corner of the world—the first of eight children born to a furnace operator in the Kingdom of Piedmont-Sardinia. Forging bricks and tiles out of the earth made for a hard life, but as a boy, there was lots of time to wonder and explore. Most Sundays and many winter days, young Schiaparelli would curl up with a book; one of his favorites was *Geography for the Use of Princes*. In the evenings, he would catch glimpses of the rings of Saturn from a telescope in the bell tower of the church.

A few weeks after Schiaparelli turned thirteen, however, the king declared war on the Austrian Empire, only to be driven back to the foot of the Piedmont. He abdicated, and his son began his reign by suppressing insurrections, revolts, and rebellions as they arose. Sardinia teetered, and Schiaparelli departed for Turin, entering university at the age of fifteen. Given all the uncertainty—for his nation, for his family—he resigned himself to becoming an engineer: a solid, useful, profitable profession. He did well on the difficult exams, which culled his class from fifty-five pupils to fifteen, and at nineteen he took a degree in civil architecture and hydraulic engineering.

There was no money left, and no scholarships, so he briefly took a position teaching elementary mathematics at a nearby school. He cut his budget to the bone. He lacked towels, he wrote to his parents in 1856 from his new post at the Gymnasium Porta Nuova, and he'd torn his best shoes walking through the meadows. He requested three or four buttons to replace his that had broken. In his shabby room, he spent his evenings engrossed in books, mostly about modern languages and astronomy, and he kept notes in a series of diaries. He wrote in both prose and poetry, in elegant cursive, alternating between Italian and his growing mastery of Latin, French, Greek, and Hebrew. He was often hungry, and that slowed him down. He found himself frustrated by how little he would accomplish by day's end and, "what is worse, in things I can't imagine ever being useful."

But then in 1857, with the help of one of his former professors, an unexpected opportunity to study astronomy came his way. He packed his meager things straightaway, overjoyed by the chance to go to Berlin to work with an expert in comets. His parents were startled and anxious, but he offered to write to them every step of the way—from Chambéry, from Paris, from Brussels, from Cologne, and finally from Berlin. "I am not going into barbarian country," he promised, assuring them that France and Germany were "civilized nations."

Upon his arrival, he plunged into the work at hand while also exploring philosophy, geography, meteorology, physics, and terrestrial magnetism. He devoured the canon of Friedrich Schiller's dramatic works, dabbled in Indology, and began learning Arabic and Sanskrit. Two years later, he traveled to Potsdam, then sailed to St. Petersburg for additional studies at the Pulkovo Observatory, where he worked under a pair of father-and-son astronomers. Soon word came of a job offer. He would begin as the *secondo astronomo*—the "second astronomer"—at the Royal Observatory in Milan. He promptly spotted a new asteroid—he called it Hesperia, "hope"—then shortly thereafter discovered that shooting stars were in fact the tails of comets.

A few years later Schiaparelli turned his attention to Mars. In the summer of 1877, from an observatory situated on the rooftop of the Brera Palace, he began jotting down notes about the peculiar features he saw. First, he cleaved the globe with a giant *diaframma,* a north-south dividing line between the darker and lighter areas. He split the bright areas into a drove of islands and gave them names like Zephyria, the home of the west wind; Argyre, a mythical island; and Elysium, the land of dead heroes. To the west, inside the Columns of Hercules, he mapped the dark seas: Mare Tyrrhenum, the sea of the Etruscans; Mare Cimmerium, the sea of the Thracians; and Mare Sirenum, the sea of sirens. To the north were the stomping grounds of Arcadia, and to the east was Solis Lacus, the lake of the sun.

Schiaparelli drew from Herodotus, the *Odyssey,* and bands of heroes in Greek mythology. Some of his names reprised moments of magnificence, like the journey of the Titan god Helios, who drove the chariot of the sun across the sky. He traced its path across Mars as he labeled the landscape, from the Orient with Chryse and Argyre—Thailand and Burma—to Margaritifer Sinus—the Pearl Coast of India. He drew from religion, borrowing names from the Bible: Alongside Arabia was Eden, large and bright. There were touches of sadness too, perhaps a holdover from the "black melancholy" he felt in his youth, or a sense of unease in the world. There were chains of smoky lands and blackish lakes. There was Memnonia, a leaden patch that sometimes whitened, but only in winter. There were also features named for places on Earth that had never been found by modern explorers, like Phison and Gehon, the lost rivers of Mesopotamia. And the whole map was upside down. His telescope inverted the image, and he found it easiest to draw and think about Mars with the same view.

It was, by his own admission, "a curious and disordered arrangement." But taken together, it evoked a human cultural history, a place intertwined with our own existence and filled with the promise that our grandest ambitions might be resurrected. He had covered Mars with beautiful names, names that would filter down through the generations.

BY THE TIME Maria reached MIT, she too was becoming a mapper. Slowly, surely. She had initially thought about a career as an astronaut, going so far as to submit an application before quickly withdrawing it. She'd wanted to have children and couldn't quite fathom leaving them behind, particularly in the wake of the *Challenger* accident. But she soon discovered that there were many ways to explore and that probes built by human hands could take her farther than a space station in low Earth orbit ever could. Spacecraft could open up possibilities far beyond human reach, throwing light onto far-distant worlds.

At the same time, she knew that the field was one of high risk and high reward—that no one sets off for the frontier with any reasonable feeling of certainty, and that space exploration lives and dies on the knife-edge of technology. She had felt the sting of that herself with *Mars Observer*. Yet, collecting data, any data, directly from Mars was a rare opportunity. And so, while the aerobraking delay had been a frustrating setback for everyone, Maria was determined to make the best of it. As luck would have it, even though the orbit was far from ideal, there were a few months when getting some mapping data was possible. Those periods fell outside the phases of aerobraking, when the spacecraft was not passing through the atmosphere, allowing the instruments to safely deploy. When they did, the spacecraft's periapsis, the point of the orbit nearest to the planet, was over the North Pole. Maria was thrilled to have the chance to test her instrument, and soon things were starting to look good for MOLA: Using the preliminary data, Maria was able to render a beautiful three-dimensional model of Mars's North Pole. She rushed the findings to press in *Science,* just in time for Christmas. It was a wonderful taste of what was to come.

But as she waited for another bite at the apple, she couldn't escape reminders of the risks. Though her polar article dominated the Mars news cycle when it came out in December 1998, NASA had launched a new mission to Mars on the same day, to study the planet's climate. Like *Pathfinder* and *Mars Global Surveyor, Mars Climate Orbiter* was a low-cost affair, the third mission designed under NASA's mantra of "faster, better, cheaper." Throughout the journey, the trajectory had required minor corrections, far more than usual, but it hadn't raised many eyebrows. It was only when the spacecraft reached Mars that mission control knew something was wrong. The signal dropped as the spacecraft arced behind Mars for the first time. But it dropped forty-nine seconds earlier than expected, and it never returned. Within half an hour, it was clear that it had hit the top of the atmosphere and disintegrated, the result of a mix-up between English and metric units in the navigational commands.

Within a couple of months, another "faster, better, cheaper" Mars mission failed. It was the same mission I'd made a tiny contribution to, with all those experiments in the Mars Surface Wind Tunnel to assess the dust loadings on solar panels. *Mars Polar Lander* nearly touched down at Ultimi Scopuli, in the south polar terrain, among canyons carved by katabatic winds and dark dunes that bloomed like flowers. But the onboard computer misinterpreted a jolt from the lander legs. Sensing touchdown, it sent a signal to shut down the descent engines. The spacecraft fell forty meters, impacting the surface at high speed.

When Maria entered the field, she knew it would be the ultimate in high-stakes science. Half the missions to Mars have failed: Some, like *Mars Observer,* sailed silently by the planet; others crashed hard, littering their wreckage across the surface. Strewn about the sands of Samara are hunks of metal and tangles of wires. West of Alpheus Colles, there's a pennant with the state emblem of the Soviet Union. The lifeless solar panels of a British astrobiology mission, one still folded like a lawn chair, gather dust on the plains of Isidis. And somewhere in Ultimi, there's an American CD-ROM, likely shattered, that once contained the names of a million schoolchildren.

AFTER *MARS OBSERVER,* Maria felt like she'd utterly failed—an above-the-fold, front page of *The New York Times* kind of failure. It was not a good feeling, but eventually she learned to draw strength from the experience: Anytime something went wrong, she would think, "How bad could this be? Worse than losing your spacecraft three days before reaching Mars? Probably not."

Mars Global Surveyor had been imperiled, with its wing dragging in the Martian night. But much to Maria's relief, the mission didn't end up as another character-building experience. Far from it. It would become one of the most successful in NASA's history. In just the first two years of mapping, it collected more data than any previous Mars mission, with MOLA leading the charge. Each day, 900,000 laser bursts were shot at the planet, and the topographical measurements they returned were superior to those we had for vast regions of the Earth. The data gave us everything: heaps of phenomena that affect the Martian surface, from volcanism and cratering and deformation to the erosional and depositional action of water and ice.

Maria and her team used the data to fill in their Technicolor map. From the sunken lowlands—a deep cerulean blue—the crust heaved forward in a riot of reds and purples, as if lava were still bursting from the planet's distended side. There were maze-like valleys and jagged fractures and mountains on pedestals. The polychromatic gash of Valles Marineris reached impossibly deep—nearly five times as deep as Africa's Great Rift Valley—before giving way to the crenulated ramparts of Noctis Labyrinthus.

Yet nowhere among the splayed lowlands or heavily cratered highlands were obvious earthquake faults or mountain belts, which would have been expected if the planet had plate tectonics. The southern hemisphere had been punched by a massive asteroid at some point in the deep past, as evidenced by an enormous depression. Hellas basin had been studied before, but no one had realized it was deep enough to swallow Kilimanjaro. On the map, it was a piercing purple-blue, like a giant round eye. It had a kilometer-and-a-half-high rampart, so it must have thrown out enough material to cover the continental United States in a three-kilometer blanket of rock. The impact might have whacked Mars hard enough to irrevocably disrupt heat flow across the core-mantle boundary, turning off its magnetic field forever, allowing the planet's atmosphere to float away.

The precision of the map allowed Maria to read the planet's history like a type of braille. As hinted by the initial data, the northern hemisphere proved to be the smoothest surface that had ever been observed in the solar system. Most of the terrain seemed to tilt slightly to the north, suggesting that a planetwide drainage system may have once emptied there, into a great northern ocean. Inscribed onto the surface was even a possible shoreline, Deuteronilus, which could be traced for thousands of kilometers. The coast ran along nearly the same elevation, with variations that could be explained by the ground rebounding, exhaling as the weight of a sea of long-gone water evaporated. With each new detail Maria plotted, another aspect of Mars's history came to life.

Mars Global Surveyor changed what it meant to see a planet. If the old map of Mars was a simple picture, the new map was a portrait. It went beyond what our eyes could take in, capturing data on contours, on composition, on forces we could not see—not just topography but things like magnetic signals and mineral compositions measured out beyond the visible wavelengths. There were subtleties to be seen—we just had to get there, and when we got there, we had to know how to look.

For years, *Mars Global Surveyor* monitored changes on Mars: the swelling of the polar caps, the ablation of rock, the disappearance of dust. And, much to Maria's delight, there were weather reports, relayed by NASA to the public through a website, a weekly window into the changing conditions of an impossibly far place, just like Pickering's missives from the central mountains of Jamaica. One day, while driving across Massachusetts, Maria even heard two voices on the radio talking about the temperatures and dusty skies up on Mars. They marveled at how one could know such things and wondered who were the scientists that had made it possible.

BEFORE THE END of the mission, *Mars Global Surveyor* also took the first photograph of Earth from another planet. To most people, it was just a footnote, anticlimactic even. It wasn't Yuri Gagarin peering through his window, seeing our planet from space for the very first time. Nor was it *Voyager,* turning its camera back from the edge of the solar system to capture Earth as little more than a pixel—in Carl Sagan's words, "a mote of dust suspended in a sunbeam." But I remember staring at that picture on my computer. I had searched for it, dragging the image to cover the screen, making it as large as I could. Earth seemed impossibly distant, but it was still recognizable. It was a crescent, more than half dark. The moon was next to us, with the same shadow.

As I peered at the pixels, trying to resolve the tiny squares of blue and green and white, I thought about where I was in the picture. In the frame, so much had been captured, and obscured. I was there somewhere, going about my life the day it was taken. I wondered if I'd been cooking a white-bean soup or sleeping in a rumpled bed when the shutter clicked, or gazing through the heavy glass window of the library at the rain outside. Perhaps I was sitting under a magnolia tree, or rushing across a crowded city street, midway through an infinitesimal step. Perhaps I was lost in thought somewhere, wondering if I had what it took to be a scientist, wondering who would show me the way.

Maria was there too, in another pixel. She was no doubt doing great things, unaware of the gravitational force she exerted on the lives of many young women like me. As I sat at my computer, right there on my screen, beneath that image of the crescent Earth, was a message signed with a quote from Tolkien. I had read it again and again: "Not all those who wander are lost." The message was the reason I'd searched for the image, the reason I was sitting there wondering where I might be the next time a shutter clicked on Mars. It was from Maria, who had written to say that in another few months, I could move to Boston to begin graduate school. She had agreed to become my PhD thesis advisor.

CHAPTER 8

The Acid Flats

THE AIR WAS filled with the smell of jacarandas as I waited at JPL visitors' gate in 2004, marveling at the shimmering posters of spacecraft. I could scarcely believe my luck as I watched the people in cars flash badges to the guards beneath a signpost saying WELCOME TO OUR UNIVERSE. Just the day before, I'd received an invitation from one of my professors to join him and two classmates in Pasadena. They'd been stationed at mission control for several weeks, working on the *Spirit* and *Opportunity* rovers, which had landed on Mars in January. Now he asked if I'd like to come to JPL for a few days, just to see what it was like.

I'd nearly tripped over myself when Maria agreed to let me go. I ran down the stairs of MIT's Building 54, pedaled past snow on the ground back to my apartment, then raced to Logan Airport to catch a flight. I had with me only my backpack: a pullover, some jeans, and a binder stuffed with notes for my first-year graduate courses in geobiology, Geological Image Interpretation, and Dynamics of Complex Systems.

The professor who'd sent me the invitation, John Grotzinger—otherwise known as "Grotz"—was one of the world's best sedimentary geologists. He'd made a well-orchestrated shift into planetary geology, just in time to explore the first sedimentary terrain on Mars. He was tan and thin and tall, and he was intensely inquisitive. When he stared at rocks, he stared like a wolf. His favorite place was among the cliffs of Oman and Namibia, tracking the rise and fall of ancient seas. He loved geology more than anyone I'd ever met. He could pick up any old shale and make it seem chock-full of possibility.

When he walked into the JPL badging office to meet me, he looked surprisingly disheveled. A "sol" on Mars is slightly longer than a day on Earth, so the science teams had been trying to sync their circadian rhythms with Mars, not Earth. As the sun set at one landing site, the rover beamed a downlink with completed measurements and end-of-drive images of its new location. Based on that data, the scientists would spend the Mars night planning the rover's next moves. As the sun rose over the rover, a new slew of commands would be beamed to an orbiting spacecraft, *Mars Global Surveyor* or *Mars Odyssey*, then relayed and caught by the rover's antenna.

Keeping pace with a distant world was not easy, Grotz explained. The rotations of Mars and Earth are similar, and living on "Mars time" just meant staying awake thirty-nine and a half minutes longer each successive day. But it was clear from his exhaustion that the small offset made a huge difference. Every eighteen days, the beginning of the day turned into the beginning of the night. Then, eighteen days later, the beginning of the night was the beginning of the day again. Mission control was slowly pulling away from Earth, then slowly ebbing back, a kind of incessant jet lag.

I thought about how strange that was—how jet lag hadn't even been a phenomenon for more than a few decades. Now we were adjusting not only to the spin of our world but to the spin of another. Grotz told me how a watchmaker in nearby Montrose had designed mechanical watches that lost a second every thirty-six seconds, which helped some of the team members keep track. *Spirit* and *Opportunity* were on exact opposite sides of the planet. *Opportunity*'s operations on the fifth floor of Building 264 began just as *Spirit*'s ended on the fourth, so as one floor of weary scientists would empty into the elevators, the other would fill. Worst was switching between the rovers, he said, which some members of the science team did periodically. That was like waking up in China.

As we walked across JPL's campus, he brought me up to date on the mission, which was designed to understand the history of water on Mars, to probe the planet's rocks and soils for clues to a warm and wet past, and to follow the water to life. The rovers were the size of golf carts, identical to each other and both far more capable than little *Sojourner* had been.

Cocooned in inflatable airbags, they had bounced to rest on opposite sides of the planet, with *Spirit* landing first, then *Opportunity* three weeks later. *Spirit* touched down in Gusev, a crater about the size of Connecticut, at the edge of the northern lowlands. Snaking its way into Gusev from the southern highlands was one of the largest channel systems on Mars, Ma'adim Vallis. Everyone hoped the landing site would be the remains of an ancient crater lake, that it would prove that there had once been huge open bodies of water on the surface of Mars, but the rover came to rest on a plain of pure lava. It was a "basalt prison," battered by dust and incessant crosswinds. On the horizon, some promising hills poked through, but it would take months to reach them.

Opportunity fared far better. It had been sent to Meridiani Planum, one of the safest places to land on Mars. It was so smooth that some of the scientists worried there would be nothing interesting to see. It too seemed promising for water, though less promising than Gusev. From orbit, *Mars Global Surveyor* had spotted an iron-oxide mineral called gray hematite, shining like a beacon. It was a crystalline form of rust, a sign that the surface might have interacted with water. It wasn't a sure bet—magnetite in volcanic lavas can transform into hematite without water being present—but it proved too beguiling to resist.

After glancing across the surface and rolling to a stop, *Opportunity* unfolded like a piece of origami. The solar panels flapped open. The paddle of the high-gain antenna tilted to the sky, and just before the wheels clicked into position, the navigational camera began snapping black-and-white pictures.

As the first image appeared on a screen at JPL, the team began clapping. But Maria's friend Steve Squyres, the Cornell geologist who was now in charge of both rovers, couldn't get his bearings. As he looked at it, he felt completely disoriented. Where were the rocks? Every last picture taken from the surface of Mars—from Chryse and Utopia to Ares Vallis and Gusev—had been strewn with rocks, filled with the kind of things a rover could study. The first image beamed back to Earth was grainy and rear-facing, but it was clear enough to make out the dents of the bounce marks in the uniform dark soil. They were the only recognizable features. Perhaps the doubters were right about Meridiani.

When the second image appeared, a forward-facing Navcam, Steve stared at it impatiently. It was underexposed, too dark to make out. As one of the engineers adjusted the contrast to make it more visible, the room grew strangely quiet.

Suddenly, tessellations of bedrock appeared out of nowhere, a jaw-dropping, beautiful wall of bedrock. The team erupted—there was laughing, cheering, crying, people jumping up and down. Steve could barely get his breath as a voice announced, "Welcome to Meridiani. I hope you enjoy your stay!" It was the first time anyone had ever seen bedrock from the surface of Mars. Sure, there were rocks at other landing sites, but those rocks could have come from anywhere. These rocks had been formed in Meridiani. And they had been formed by water.

They looked just like water-lain sedimentary rocks—in other words, the kind of rocks that tell a clear geological story. For Steve, it was too good to be true. He had advocated for Meridiani over a hundred other sites because he hoped to find evidence of water, but in his heart, he never dared hope that he would actually discover what he was looking for. He was almost afraid to believe it. He stumbled to the front of the room and said, "I will attempt no scientific analysis . . . holy smokes, I'm sorry, I'm just, I'm blown away by this." Someone yelled, "Did we hit the sweet spot?" Steve stammered, "The sweetest spot I've ever seen."

Up until that moment, no surface mission had ever observed water-lain deposits on Mars; there had been no stratigraphy to investigate—no layers, no relationships among the layers—and therefore no way to investigate how the geology and the climate and Mars might have changed over time. Sedimentary rocks that we could poke and probe were the holy grail. Finally, we could really peer back through time.

As other Navcam images came down, however, the horizon seemed strangely close, and it was impossible to get a sense of scale. The team tried looking at the images from different angles, struggling to figure out exactly what they were seeing. Slowly, it dawned on them that the airbags must have rolled to rest in the depression of a small impact crater. A crack about the interplanetary "hole in one" led to the crater's name—"Eagle," for two strokes under par. And then, when *Opportunity* finally kicked into gear and began to approach the bedrock, the team's perspective shifted again: The feature they had called the "Great Wall" became "Opportunity Ledge." As it turned out, it was barely ankle-high.

JUST AS MARS *Global Surveyor* had predicted, there was hematite everywhere at Meridiani, another line of evidence for water, but not in a form anyone had expected. The hematite had collected on the surface like a spray of ball bearings. The pictures of the ground looked like the kind of cartoonish surface Fred Flintstone might slip on. The spherules were gray-blue in the color-stretched images, so they were dubbed "blueberries." It was a perfect name—they were also spread throughout the rock just like blueberries in a blueberry muffin.

At first, the team wondered if the "freaky little hematite balls" were raindrops of metal that had erupted volcanically and been lobbed into the freezing air, solidifying mid-flight before falling back to the surface. But they weren't all in a single layer, like a layer of ash; many were also buried in subsurface sediments and evenly spread apart. And if they were made by a volcano, where was the volcano? Then another idea emerged: perhaps they were concretions, little metal balls that had swelled from a point of hematite in the subsurface. Slowly, more hematite could have globbed on, forming layers upon layers, like pearls inside oysters.

The infrared signatures also argued against a volcanic origin, instead suggesting that the hematite had formed in the presence of cool percolating groundwater. The blueberries strewn across the landing site were small, the size of peppercorns, and they were relatively uniform, indicating that the same amount of groundwater was probably present for the same length of time throughout the region. After the water table receded, wind continued to beat away at the surface, eroding the soft surrounding rock. Slowly, the blueberries would have fallen out of the rock and rolled onto the ground.

The bedrock and blueberries weren't *Opportunity*'s only early discoveries: There was also magnesium sulfate everywhere, like Epsom salts in a bath, stretching in every direction, likely deposited in a lake or shallow sea. And there was another kind of sulfate, one that was even more surprising. Not long after landing, *Opportunity* had driven up to a rock called "El Capitan," drilling into the rock's interior with its tiny grinding wheels. Looking at the data, the team saw what they believed to be the sulfate mineral jarosite. Jarosite indicated highly acidic conditions, which, team members were quick to point out, didn't exclude the possibility of life—microbes, after all, survived in acidic waters in places like the mines of the Sierra Almagrera and the Río Tinto, Spain's "river of fire." Moreover, jarosite was a hydrated mineral, meaning it simply couldn't have formed without the presence of water.

There was evidence for standing water too: layers of rock overlapping and cutting into others in distinctive patterns—petrified horizons of sand and sediment that had once washed over one another in shallow rivulets—the same smile-shaped markings that line the bottom of most streambeds on Earth. There had once been salty seas or lakes and streams, with sediments reworked by wind. This was what the Mars science community had been sent to find, the first definitive evidence for liquid water on another planet. When the team looked closely at the outcrops of Meridiani Planum, they could even see ripples in the soft rock.

. . .

IT TOOK A moment for my eyes to adjust after the elevator opened in Building 264 and I followed Grotz into mission control. The windows were sheathed in thick black vinyl. There in the sunless room, dissociated from time, two dozen scientists sat in high-back blue chairs clustered around tables of computers. Giant screens shone with running clocks, projecting the details of the sol's timeline. In the middle was a long sleek desk, stretching some three to four meters in length, covered by stacks of orbital images of the landing site. The room felt like the helm of a frigate, sailing through a darkened ocean.

As I looked around, I saw dozens of scientists I knew, mostly only by name. Then I saw my undergraduate mentor Ray Arvidson, who grinned and asked, "How you doing, kid?" A moment later, Steve strode to the front of the room in a pair of cowboy boots. He tossed up a microphone and caught it again. "Time for 'sog,'" he said. Everyone rose from their seats as I tried to figure out what he meant. SOWG, I would quickly learn, was the Science Operations Working Group, the most important meeting of the day. Still holding my bookbag, I followed the others to a nearby room. In the hallway, there was a bottomless freezer of frozen treats. The gift of a local dessert distributor, Ray told me, tossing me a chocolate Drumstick.

As the mission's second-in-command, Ray took a seat behind a little blue placard. He was SOWG chair, the head of the meeting for the day. I slipped into a seat at the back and scanned the room. People were talking about "yestersol," "solmorrow," and the upcoming "soliday." It was at once so childlike—all that dripping ice cream!—and at the same time so full of technical sophistication. The rover's next measurements were designed to follow up on the discovery of the blueberries—critical to piecing together the history of water on Mars. The scientists and engineers worked alongside one another for more than an hour, hashing out priorities for the drive and figuring out how to implement them: which specific instructions would be sent to the rover through the Deep Space Network. To the uninitiated, they might as well have been talking in code.

But then the meeting ended, and the lightheartedness instantly returned. The team was bonded and always orchestrating well-meaning pranks. One morning the MiniTES team—the scientists manning the Miniature Thermal Emission Spectrometer on the rover's arm—had arrived to find their computers and chairs swaddled in Saran wrap. Then the following day, the Pancam team, which ran the panoramic camera on the rover's mast, discovered that all the keys had been popped off their keyboards, save for one line of letters: "M-I-N-I-T-E-S."

OPPORTUNITY WAS ON its way to Endurance Crater, and Grotz suggested I stay until the rover arrived. I took up residence on my friend's Caltech couch, taking taxis to and from JPL, as each sol brought the rover closer and closer to its destination. The crater was as large as a football field, and the team had been eyeing it on the horizon since the early days of the mission. *Opportunity* was proceeding from one hole that had been punched into the ground to another. It wasn't the craters themselves that were the jackpot but the crater walls, for they offered a glimpse into the ancient past. They revealed layers that had been stacked like the pages of a closed book, one moment in time pressed close against the next.

After leaving El Capitan, the rover had stopped at a sinuous crack named Anatolia, then a small impact crater named Fram. The terrain had become sandier on the route to Endurance. Each day the engineers had to make hazard maps for the rover, and the rover could only go that far.

On Sol 94, a scientist named Tim Parker called out from the far side of the room, "First Navcam frame is down. . . . Anybody want a look?" We gathered around his computer, and as he opened the file, there was a collective audible gasp. The rover had climbed out of its landing crater many sols ago, traversed farther across Mars than any vehicle before it, and was now lurching over the chasm of Endurance. Its two front wheels were beyond the lip of the crater, which had been three meters closer than the engineers realized. *Opportunity* had just driven and driven until a hazard warning kicked off the power.

He threw the image up on the huge screen in the front of the room, and we slowly fell back into our chairs. Ours were the first human eyes to peer into that mysterious abyss, and it was one of the most breathtaking things I'd ever seen. As I stared into the center of the crater, I felt like Alice in Wonderland falling through a rabbit hole. "What is this world?" I thought, there on the verge of Endurance, my eyes wide. "What is this piercingly wild place?" The giant cavity was laced with hummocks of sand. The most ethereal gossamer dunes filled the void at its center, unlike any dunes I'd ever seen. They looked like egg whites whipped into soft pinnacles. And enveloping the edges, there was undulating outcrop, cut with gorgeous striations, deeper than I was tall. I was supposed to fly back to Boston in a couple of days, but this I couldn't leave.

I decided then and there that I'd do whatever it took to convince the team to let me stay. I had never wanted anything so badly. I lingered at JPL that night, even after the sol completed, trying to figure out how to make it happen. I could take incompletes in my classes, I realized; I could sleep on my friend's couch, on the floor if I had to. I could help manage the flash data storage, completing the daily checks. I could train as a PUL, a Payload Uplink Lead, uplinking code to the microscopic imager. I implored my mentors, and when they agreed to take the request to Steve to consider, I nearly leapt from my chair.

At the time, I wasn't sure if Steve, as the leader of the entire mission, even knew my name, and I had no idea how he would respond. But as it turned out, Steve's own path into planetary science had been rather serendipitous too. It started with a high school friend falling from a pair of ice axes. Steve was a teenager learning to ice-climb, trying to shore up his mountaineering skills to be admitted to the Juneau Icefield Research Program, a summer program funded by the National Science Foundation. He'd accompanied his friend out to western Massachusetts, and when his friend's wrist cracked apart, Steve had to splint it and carry him to safety. His friend's mom wrote him a recommendation for the program, which clinched his admission.

So Steve spent the summer of 1974 skiing, a pair of red gaiters flashing across the ice. It was there that he fell in love with vast and distant places. He initially thought he might go into marine geology and spend his life at sea. Perhaps he could map the bottom of the ocean, a colossal frontier that had never been surveyed. Then one afternoon as an undergraduate, he saw a new map posted in the geology department. It showed mid-ocean ridges zippering across the globe, all in intricate detail. "Well, crap," Steve thought, "the ocean's been done." The next semester, on a lark, he signed up for a graduate course on the *Viking* mission results. The final assignment was to analyze the original surface images. He swung by the "Mars Room" late one afternoon, expecting to pick a topic in fifteen to twenty minutes. Four hours later, he emerged knowing what he wanted to do for the rest of his life.

There weren't many places in the United States to get a PhD in planetary science, to study geology but not as it occurred on Earth. Cornell, where he was already an undergraduate, was one. He applied "for completeness," expecting to make his way to a new university, with no inkling that he'd spend most of his life in Ithaca. A few weeks later, he received a note from Carl Sagan, who'd seen his application, suggesting the possibility of working with him on the *Voyager* mission. Sagan was a Cornell professor, but he spent most of his time in California. They hadn't ever met, but what could be more exciting than the "Grand Tour," two tiny probes that would travel for the first time to all the planets of the outer solar system? Steve accepted Cornell's admission offer. Even though Sagan barely set foot in Ithaca while Steve was in graduate school, they would see each other at JPL often. At the time, Steve had no idea how much planning went into space missions—the engineering, the rocketry, all the preparations, all the cost. He was a student, and when he showed up and was provided with the data on a platter, he didn't have to ask those kinds of questions. He felt like he'd been given the keys to the universe.

And that's how I felt when Grotz told me I could stay, when Ray offered to arrange for a badge, when Maria assured me she would clear any bureaucratic hurdles with my graduate program. "This is how it begins," I thought, "with a bit of luck!" I didn't appreciate it at the time, but Steve had spent years learning every circuit on the rovers, every pyro, every destination of every bit of wire. He'd met every engineer and considered every failure path. I'd done nothing to help get those rovers to Mars, but because of his hard work, I'd been given the chance to zoom to the very epicenter of modern planetary science, where everything was shot through with light. Instead of packing for my flight home, I called the airline. I couldn't stop smiling when the woman on the phone confirmed, "Return indefinite."

THERE WAS NO shortage of fascinating work. As we inched below the western ridge of Endurance Crater, we realized the rocks were stippled with blueberries all the way down. And as we looked at orbital images, it seemed that the same geologic formation probably extended tens if not hundreds of meters down and kilometers across. Not a little water but a *lot* of water would have been required to make them.

On the opposite side of the planet, *Spirit* was heading for the Columbia Hills, which rose like islands above a sea of incessant basalt. We reasoned there might be sedimentary rocks there, once deposited in water—rocks that hadn't been buried by lava. The rover made great time, covering real distance. There had never been a mission like it, and every day there was a new vista. When *Spirit* reached the base of the West Spur of Husband Hill, over three kilometers away from its landing site, it uncovered traces of hematite, along with chemical enrichments from flowing water. And high above, tens of meters up the hill, there were even signs of layering.

I worked mainly on helping to document the physical and chemical changes that had occurred as the playa lakes receded, as sediment had turned to rock, tracking things like the size and distribution of the blueberries. A couple of months after I arrived, at the beginning of the summer, Grotz returned to his fieldwork in Namibia. At that point, I joined one of his graduate students in the cavernous house he'd been staying in as a Caltech visiting professor. I had never lived in such a large, bare space. It felt like the house had been occupied, then emptied, just like the tabular crystal molds in the rocks I'd begun studying. They were called "vugs." The tiny cavities—another diagenetic feature—were once filled with minerals like gypsum, which later dissolved away, leaving an angular, empty sliver behind. We knew that Meridiani was once wet, but those tiny vugs and blueberries were crucial for understanding the sequence and timing of water saturation; they were how we would understand if the waters had receded and returned and how long the wet conditions may have lasted.

We began to find little clues, like a hematite blueberry forming in the cavity of a vug, which suggested that the rock in front of us, all those billions of years ago, had already been marked by a tiny absence—that the blueberries must have formed *after* both the creation and dissolution of the crystals. Soon there was enough evidence to conclude that the blueberries were one of the last features to form and that they'd formed under distinct environmental conditions. The story was coming together: Meridiani was a place where the water table had risen repeatedly in the ancient past, where waters had drenched the surface again and again.

I'd brought next to nothing with me to California, and I'd acquired next to nothing since my arrival except for a badge and security token, a couple of books I'd found at a used-book store, and a rental car I would drive out onto the expressway to get to JPL. I'd exit and follow Oak Grove Drive as it curled north into the arroyo. Sometimes in the early-morning light, the roads felt desolate. There were no joggers running at the Rose Bowl, no students mingling on the balconies of La Cañada High School, no equestrians jumping at the Flintridge Riding Club. Even though the mission had shifted over to a modified version of Mars time, condensing the planning cycle in order to block out the hours in the middle of the night, it still felt as if the outside world had vanished. As I entered the gates of JPL, I felt like Nikos Kazantzakis, the giant of Greek literature, arriving at the wild and holy monasteries of Mount Athos. I walked beneath the olive and oak trees, then into the dark and hallowed halls. I took my place among my colleagues, at a computer with a terminal window open. Each screen filled with lines of code—commands that would beam from the mission team all the way to Mars, commands that would rebuild data files and assemble images faster than any human mind and, in so doing, illuminate the truths of a distant desert world. The scripts seemed incandescent, rasters of white letters and numbers gleaming from the black background. All the zeros were hemisected, like lines of gilded text. It felt holy to be in those rooms, committed more fully to the mission than I'd ever been committed to anything. Time had been obliterated. What remained was only the larger thing, in prismatic focus.

IT CAUGHT ME off guard one day to be invited to see the Dodgers play. Steve was opening a game, and my name, along with a couple of others, had been pulled from a hat to join him. As I walked into the stadium, I was given a blue sticker in the shape of a baseball, with big letters spelling out the words "pregame guest." I affixed mine to my jacket and followed Steve to the field. We'd brought along a handheld model of the rover, and I showed it to some of the players as the stadium seats began to fill.

Just before the national anthem, one of the officials gestured that it was time for us to make our way to the pitcher's mound. The grass was short and green and seemed to spring beneath my feet. My head arced slowly as I paced across the field, trying to take it all in. There were flags flying and, in the distance, palm trees. I tripped ever so slightly where the grass ended and the dirt began, then righted myself next to the pitcher's mound, smack in the center of the enormous stadium.

As I stood there, the announcer described the mission over the loudspeaker, booming stats about its success. An artist's animation of the rover landing played across the JumboTron, just above an ad for the restaurant chain Carl's Jr. As the stadium began to roar with applause, everything seemed to slow down. I felt the heat of the brown dust I'd kicked over my sandal, its warm grit gently falling between my toes. I realized I was supposed to be waving, but I was standing there motionless, unable to take my eyes off the teeming crowd. Here we were, a species. Thousands of human lives, thousands of tiny bodies in rows up to the sky.

I drove to JPL later that night, swiped my badge in the turnstile, and roamed the campus until I found myself in Building 264, sitting alone in one of the high-back blue chairs. My mind kept wandering to the little tribe of scientists on the mound, standing there in the sunny stadium but mentally still far, far away. I realized that as I had become more and more absorbed in the mission, the fabric of connection that comprised my life in Boston had started to fade away. I no longer did ordinary things, like going to the bank or shopping for socks. I talked to my friends and family less and less. Staring up at those fans, though, I was suddenly, jarringly back in the middle of the world I had left behind, a human world. I wondered if Lowell had felt the same way, stepping away from his telescope, returning from the thirsty Arizona desert to Sevenels. Or Dollfus, thudding down from the sky into a cow pasture.

In the grandest sense, we'd made it to Mars. If space were a giant sea, we'd found the next atoll. We'd fluttered out into a darkness of scattered islands, and we'd made landfall. We were now exploring like a restless finch, hopping among the rocks.

We could even see our tracks from orbit—ever so faintly. They'd been pressed into the frigid soils, right up to the edge and down the side of Endurance. I thought about Ernest Shackleton's ship, for which the crater had been named. It had been caught in the ice, crushed in the cold. As I looked around at the room, I saw the signs detailing prevention strategies for ANTICIPATIVE FATIGUE. Eat smaller meals. Drink lots of water. Try to exercise but never before bed. I saw the empty chairs, the motionless piles of papers.

There was something I'd put on a shelf in my mind since the early days of the mission. It had all been so exhilarating at first—for years, NASA's Mars strategy had been "follow the water," and we'd found it, thousands of swimming pools' worth of water. Water that had inundated the surface, water that had collected in playa lakes, reflecting the sky.

But I couldn't stop thinking about the jarosite that *Opportunity* had found in El Capitan. All the jarosite we knew of, in all the amber-yellow cracks and crevices where it was found on our planet, was formed in acidic water. And not just acidic water but highly acidic water. The U.S. Postal Service refuses to let scientists ship samples of it through the mail—it's that corrosive.

In places likes the ephemeral playa lakes of the Yilgarn Craton in western Australia, microbes can survive in pHs less than 2, right alongside jarosite. But it's unlikely that life on Earth sprang from waters this acidic. Tiny organisms, even eukaryotes, can now tolerate the conditions, but they've adapted to do so, utilizing a sophisticated evolutionary machinery. Minerals like jarosite essentially form in sulfuric acid. Could we really expect life to form in these conditions?

The team had also been doing work on salinity at Meridiani, and I knew a paper was under way for *Science,* one that concluded that salt levels may also have been much too high for even the hardiest microbes. It was the same reason the soy sauce in the JPL cafeteria never spoiled. Soy sauce contains lots of water, yet not enough unbound water for microbes to grow.

How quickly chemistry could change things. We'd found water on Mars—the primary goal of the mission—and in the collective excitement, I was slow to realize that not all water gives life. The water on Mars might have been deadly.

NEWS OF WORLD events had mostly come and gone during those secluded months at JPL. I was aware of happenings outside the mission—the CIA admitting there were no weapons of mass destruction in Iraq, the ongoing saga at Abu Ghraib—but if anything, these events just made Mars feel like a place where thoughts stayed safely within the realm of geologic time.

Then one evening at mission control, I heard some of the team members talking about the Olympics. Ten thousand athletes were gathering in Athens for the 2004 games, back in the country where it had all begun. As a kind of commemoration, one of the team members suggested that we use *Spirit*'s Rock Abrasion Tool to carve the Olympics symbol into Martian rock. We'd already been using it to etch out patterns, incising five-centimeter circles into the Martian rock to peer beneath the dust and rock rind, often in clusters to accommodate the large field of view of some of the instruments. We'd left them strewn about the surface. We called the clusters "daisies," as they looked like flowers: one central circle, flanked by a circle of circles.

Sometimes the daisies were dark against the bright dust. Other times they were lighter than the rock. They were instrumental, of course, necessary to enable our measurements. But they were also reminiscent, however inadvertently, of the impossibly fecund world the rovers had left behind. So one day in August, we decided to pause our normal brushing protocol at five circles on a rock named Clovis, three in a row, then two below, each interlaced. The rover's camera captured an image of the rings, the first human symbol drawn onto the face of another planet.

It reminded me of something I'd read about an Austrian astronomer proposing in the early 1800s that giant trenches be dug in the sands of the Sahara. They were to be in the shape of mathematical symbols, filled with kerosene and set alight in the desert's darkness with the hope that the blaze might be visible from Mars. And of the German, not long before, who reportedly suggested sowing wheat into a giant right triangle in the Siberian tundra, bordered on each side by a square of pine forest, thereby invoking the Pythagorean theorem from space, which, of course, "any fool would understand." And the Frenchman, not long after, and his network of seven mirrors that, if arranged properly across Europe, might beam up the shape of the Big Dipper.

How tender it would have been: fires in the night, a field of pine, a glint of light. How tender and how futile. And now we too had been part of this, except it was Mars we marked, inscribing our ephemerality into terra firma, into one of the driest, oldest rocks onto which anything had ever been inscribed.

In a way, the rings were a beautiful gesture. They felt as spare and evocative as the petroglyphs in Damaraland, pressed into the ochre walls of Twyfelfontein: an antelope on the plains, a shot bow. And in the Libyan Desert, the Cave of Swimmers in Gilf Kebir: bodies arced and gliding through an ancient lake, now vanished and buried beneath the sands of the Sahara.

At the same time, what if it was pointless—the daisies, the rings, the fact that athletes in Athens were flinging sticks, running fast, and diving into the cool water that surrounds our planet? Would any of it matter against the backdrop of an empty cosmos?

. . .

When summer turned to fall, I finally returned to Boston. I watched the skyline come into focus as the plane neared Logan Airport, all the tiny buildings. After landing, I walked out of the terminal and found a taxi. "No luggage?" the driver asked.

As we sped along Storrow Drive, I rested my head against the window, watching the sailboats bob in the river. When the taxi dropped me in front of my apartment, just off Beacon Street, it looked the same as it always had. I walked up the two flights of stairs and let myself in. There in my bedroom, on the windowsill, I noticed the mug of tea I'd made for myself the morning before I left. I'd forgotten it there, and it was as if it'd been fossilized. The liquid had vanished, and when I lifted the thread, the featherweight tea bag detached effortlessly from the ceramic. I stared at it for a long time, wondering what it said about me that I could step out of my life so easily, and what it would mean to step back in.

PART 3

A Boundary Is That Which Is an Extremity of Anything.

—EUCLID'S ELEMENTS

CHAPTER 9

In Aeternum

OUT IN SIBERIA, the Kolyma Lowland stretches across the far eastern steppe. The gray sky hovers close to the sediments, breeze-blown silts cemented by massive amounts of frozen water. Great fists of land are upturned into hummocks by expanding and contracting ice. The Kolyma Highway ebbs through it, the "Road of Bones," named after the thousands of gulag prisoners who died constructing it. Human skulls were once so common that children used them to gather blueberries.

In the 1990s, shivering scientists had begun to drill cores along a twelve-hundred-meter stretch between the Lena and Kolyma rivers. Deep below the thawing surface, they accessed places where the temperatures never warmed, where there were no water-bearing horizons, where thick ice veins had persisted intact for a million years or more. They were careful to control for contamination—their own human cells, the microbes that lined their tools, even windblown bacteria. They slowly, painstakingly, rotated a corer to avoid using drilling fluid—twenty, thirty meters deep. They painted the cores with cells of *Serratia marcescens* as a marker for contamination. Then they carved away the outermost surface with sterile scalpels and sealed the pared-down batons of frozen earth within the sterile metal cases of a borehole freezer, hauled out to the middle of nowhere.

Dice-sized cubes of those cores eventually made their way to laboratories in Oxford and Copenhagen, where some of the best ancient DNA research in the world was being conducted. They were carefully peeled open, tiny rootlets snapping as they were exposed to the air—the ancient roots of ancient plants, still gripping grains of soil. Within those same frozen lumps of permafrost, the DNA of woolly mammoths, lemmings, and reindeer was uncovered, as well as spores, pollen, and the traces of ancient microbes—all perfectly preserved. These were the remains of an ice-age ecosystem, one that had been lost to the world.

After leaving JPL, I'd read about the study, transfixed by the exquisite resolution of data, the resilience of those delicate molecular fingerprints. In particular, I thought about the microbes. What happened when life was pushed to its very limits? How long had they survived in the subsurface? If life could persist underneath former western Beringia, I reasoned, perhaps life could persist on Mars too. I knew there was other water on Mars—frozen water, swaths of it laced underground. Not just at the poles but all the way down to the low latitudes, and it had been there for millions of years. To many people, the most exciting thing about that discovery—made in 2001 by a NASA orbiter—was that human explorers might one day mine that water: thawing the ice, baking it from hydrated minerals, using it as a resource in colonization. But I couldn't stop thinking about whether life could be embedded deep in that frozen ground, beyond the reach of the acidic waters.

IT WAS ONLY a few months later, in the dead of winter, that I found myself at the Niels Bohr Institute in Denmark. I'd proposed to Maria that some time in Europe would expose me to cutting-edge biology, and she gamely agreed to another prolonged leave from campus. When I arrived in Copenhagen, I moved into a pocket of a room on the ground floor of a laboratory building. There was just a chair, bed, desk, and window looking out onto an empty greenhouse. It reminded me of the miniature dioramas I'd made inside shoeboxes as a child.

I was in a country of strangers, and everything seemed quiet. Parents left their babies in carriages outside store windows, their sleeping bodies bundled against the cold, their faces tucked below a billowy sleeve of fabric. Danish cars glided through the Danish snow. Sun glinted off iced-over bicycles. Even the restaurants and coffee shops, with their white walls and beautiful dishware, seemed hushed.

I'd come to Denmark to learn about ancient cells and ancient DNA, to see if cells could persist in ancient permafrost, in some of the most challenging conditions on Earth, to help figure out whether the microbial DNA recovered in those Siberian cores was DNA from living cells—potentially, possibly, surviving over ancient timescales—or simply the remains of creatures long dead, like the woolly mammoth.

I had the chance to work with one of the pioneers in the field, Eske Willerslev, as he was setting up his newly formed group. I'd met him during his research fellowship at Oxford, where I'd also spent time studying. You wouldn't guess it from his chain-smoking or the black-and-white-checked Vans on his feet, but he had once disappeared into the wilderness for nearly four years, living part of the time as a fur trapper. He survived off moose meat as he and his twin brother canoed up untamed rivers and explored the frontier tundra. He was a brilliant scientist but also a bit of a nomad. Like me, it seemed, he traveled light.

Most days, I alternated between two labs, one in the building where I was living, and the other in an older structure across the way. The labs had been separated to isolate the cleanroom. Contamination was such a pervasive issue that the cleanroom had not only been built in a different building but also secured behind a pressurized antechamber. If there was life at all in that permafrost, we knew there wouldn't be much, and we had to be very careful. I never entered if I hadn't showered just before and changed into freshly laundered clothes. Inside, I worked in double gloves, sterile sleeves, a puffy Tyvek bodysuit, and a face mask. Crossing into it felt like crossing a threshold into a spaceship.

I liked my newfound proximity to the lab bench, accessible at any hour, any day. I found solace in the supreme isolation, the quiet. I hardly ever left the lab, but when I did, it was to drift through the city. I once wandered into a cemetery in the heart of Nørrebro, where the rows of headstones were desolate. Patches of snow clung to the epithet on Søren Kierkegaard's grave: "In a little while, / I shall have won, / The entire battle / Will at once be done. / Then I may rest / In halls of roses / and unceasingly, / and unceasingly / speak with my Jesus." I returned to my little room and spent the evening with an English copy of *Fear and Trembling,* listening to the whir of the generators.

After a few months of steady, constant effort, I started to see thrilling results: evidence of intact ancient cells. I'd been hoping the work would have implications for how we think about life detection, and it looked as if we'd discovered that tiny bacteria could survive over epic timescales: 300,000 years, 400,000 years, perhaps even 600,000 years.

Eske underscored the importance of proceeding cautiously. We arranged to transport a portion of the original samples to a lab in Australia, to ensure they could replicate what I was finding. But if those cells were indeed alive, *how* were they alive? How did they make it in a place so devoid of warmth or nutrients? How could they endure the ravages of time? I thought about it as I paced the city, as I ate granola in the mornings. Perhaps they shut down operations entirely, hunkering down and surviving in a state of dormancy? Or perhaps they found a way to repair the inevitable damage?

A couple of weeks later, we took the train to Lund, Sweden. In collaboration with colleagues there, we set up an experiment in small stainless-steel chambers. The cells would sit at subzero temperatures for nine months while tiny sensors attempted to measure minuscule puffs of gas as they "exhaled." If the cells were in fact "breathing," if those data aligned with our genomic results, then we would have a big discovery on our hands. It felt like a long time to wait, but of course it was just a heartbeat in the life of those cells.

In the weeks that followed, I continued to gather additional measurements. Within the building's massive stone walls, I could feel the warmth of Denmark's near-Arctic light as it fell through the windows onto my masked face and gloved hands, the heat of our enduring star. I rechecked all our positive and negative controls and ran some auxiliary experiments in the cleanroom. Late in the afternoons, as the sun began to set, I'd duck out for a tea break, pacing the halls holding my warm mug, thinking about how this place was home to some of the most remarkable contributions humans had ever made to science. The basic structure of the atom had emerged from those very rooms. Later, the Institute had borne witness to the transition from classical to quantum mechanics, to the world veering from deterministic to forever probabilistic. It was home to a vibrant intellectual history, housing its memories within sturdy walls—walls that were full of fossils, the calcified husks of lives long gone. I was there for only a moment, but in that moment I was staring into the ice age, and I was surrounded by things that were built to endure.

While I was waiting for the Swedish results, I moved back to Boston and into a new apartment. I passed my generals—the qualifying exams for my PhD—and began the intimidating work of outlining my dissertation. I also worked on the DNA paper, and each time I did, it was like taking a break from my future to look back in time. I'd think about it as I jogged through Mount Auburn Cemetery and rode on the rumbling subway beneath the city. I wondered if the cells in those tiny incubation chambers were still alive, releasing tiny whiffs of gas into the stainless-steel tubes. I worried our apparatus might not be sensitive enough to see them.

WHEN THE RESULTS came in—confirming at long last that the microbes were alive, barely but clearly respiring—I nearly burst with excitement, rushing to Maria's office to share the news. I spent the next couple of months writing up the paper. After it was submitted, there was more waiting—for the reviews to come back, then for the editors to respond to our responses. The weeks and months passed. I'd fallen into a rhythm, working as steadily as the scullers who rowed along the Charles below my office window. Before I knew it, it had been more than two years since I'd returned to Cambridge. The migrating geese had come and gone—north in the spring, south in the fall. The tree branches had held the weight of snow through the winter and then grown anew.

One crisp autumn evening on my way back from the lab, I stopped by a bar in Harvard Square. A gathering of alums from a fellowship I'd received as an undergraduate was under way, and I'd promised a friend I'd stop by. I was wearing a backpack over my coat, the kind of thing you could only get away with in a college town. As I lifted an appetizer from a tray, a charismatic guy at the center of the room caught my attention. I noticed that his "HELLO" name tag listed the same class year as me, and he was talking to others I knew. It was puzzling. I didn't recognize him, but it seemed impossible that I would have forgotten someone in my cohort. When I walked over to inquire, he laughed at my cross-examination. He'd just stopped by with a friend, who instructed him to scribble something to get a free drink. He glanced down at my shoulders and smiled at the backpack. I let it slide off my back as we talked for a few minutes.

He'd grown up in Boulder, Colorado, he told me. He'd started off in philosophy but was now studying to become a public-interest lawyer. I mentioned some of my work on Mars and life in the permafrost. I told him proudly about my new paper—my first as the "first author"—which was just going to press. I told him that I had demonstrated how a leading theory for cell survival—dormancy—came up short. Rather, the oldest cells displayed an amazing ability to slow down cellular activity and repair their genomes. They didn't succumb, I said; they recovered, slowly but surely fixing the damage that had been done.

Before I walked into the cold that evening, I looked back at him, and he looked at me and smiled. Then the wind blew the door closed behind me, and I headed home through the fallen leaves.

THE QUESTION OF survival was a live one on Mars too. Might evidence of an ancient ecosystem linger there? Might ice in the polar

Martian soil potentially become unfrozen and available to microbial cells? Might life persist between thaws? Whether extinct or extant, it stood to reason that ice and subsurface permafrost would be an ideal place to search for evidence of life. Eske's work had also caught the attention of a Mars scientist named Peter Smith, whose thoughts careened toward Mars when he realized that the entire phylogenetic tree of life had been reconstructed from those samples from the Kolyma lowlands, the very same samples I had worked with.

Smith was born the same year as my father and had the same shock of white hair. Not long after ringing in the new millennium, he decided to propose a mission that would land on Martian polar terrain to assess its biological potential. He envisioned the mission, which he named *Phoenix,* rising from the ashes, like the mythical Arabian bird, recouping the lost science from *Mars Polar Lander*. The spacecraft would be cobbled together from spare parts, with recycled hardware and software, on just a shoestring budget. To Smith's delight, it was selected as NASA's very first *Scout* mission, a cheaper supplement to the core missions of NASA's Mars Exploration Program.

Phoenix was a precarious concept right from the start. It would become the first Mars mission ever to be run and operated out of a university—the University of Arizona—rather than a NASA center. Pulse thrusters were going to be used for landing, not the inflatable airbags that had safely cocooned *Pathfinder* and the *Spirit* and *Opportunity* rovers. And somehow it was going to operate on the polar terrain, whereas almost all the successful missions to Mars had touched down within a relatively tight band around the equator. The limitation on solar energy would place severe constraints on the amount of power available for the instruments.

Smith suggested landing on the northern polar plains, a Mars latitude similar to that of Canada's Northwest Territories, not the doomed southern polar plains, where *Mars Polar Lander* had crashed. It was too dangerous to land on the pole itself, on solid ice—but at least the low elevation of the northern half of Mars would mean the spacecraft would have more atmosphere to slow it down, making for a safer landing. The targeted area in Vastitas Borealis was one of the cold, mysterious places that had just been identified by the *Mars Odyssey* mission as having extremely high amounts of water ice in the permafrost just beneath the surface.

In the spring of 2008, as the *Phoenix* lander approached Mars, a couple of hundred scientists and engineers filled two operations buildings on the outskirts of Tucson. The structures were single-story and had been dirt-brown until the University of Arizona's art students painted a fiery mural on the south wall. Meanwhile, Smith left for Pasadena to prepare for the press conference at JPL. He'd been told over and over to expect the mission to fail, so when he arrived, he stood in the von Kármán auditorium, at the same podium as all his predecessors who had led Mars missions, rehearsing a press conference revealing that the mission had crashed into the surface. The press office prepared press releases for a half dozen doomed scenarios, such as the parachute failing to open. The days unfolded one by one, with tension presumably similar to what he'd felt as a child when his virologist father became one of the first people to ever receive the vaccine for yellow fever (it was self-administered). "Well, Dad, how are you feeling today?" Smith had asked him, day after day.

The landing was scheduled for Memorial Day weekend. On a whim, the media team at JPL decided to set up a Twitter account. They reasoned that a feed would allow NASA junkies to follow the events with their cellphones. Twitter was relatively new, and at first it wasn't obvious that anyone would sign up. Veronica McGregor, JPL's media-relations manager, elected to have an email sent to her with every new subscriber. She also realized with her first tweet that using "I" instead of "the spacecraft" or "*Mars Phoenix*" would help her stay within the 140-character limit. It was a small adjustment, largely inadvertent, but it humanized the little lander. As Memorial Day neared, and the spacecraft's plucky new account got mentioned in an online *Wired* article, McGregor's computer started dinging like "a Vegas slot machine."

When the landing day came, a high-resolution camera on the *Mars Reconnaissance Orbiter* captured a remarkable photograph of *Phoenix* plummeting toward the pole. The parachute opened a full six and a half seconds later than expected. It pushed *Phoenix* to the very edge of its landing ellipse, yet the retro rockets fired flawlessly, placing the spacecraft down perfectly level, in front of a giant crater. The lander did a pirouette as it opened, aligning its solar panels in an east-west direction to maximize the light hitting them. @MarsPhoenix tweeted, "Cheers! Tears!! I'm here!"

THERE WERE MULTIPLE cameras on *Phoenix,* the smallest able to resolve the surface structure of a grain of sand, and when they flipped on, the vista was incredible. The polar terrain stretched without end and looked like one of my grandmother's well-loved old quilts. The patchwork ground had been broken into intersecting polygons, some a couple of meters in size. It was a gorgeous worn geometry, formed by the repeated expansion and contraction of subsurface ice. The team knew that their size was determined by the distance from the ice table. Some crinkling of the surface—shallow knolls and dips—had been seen from orbit, but when *Phoenix* landed, they saw "polygons within polygons within polygons," all formed under different climatic conditions. It meant the periglacial environment on Mars was complicated and remarkably active.

After a few sols, *Phoenix* extended its robotic arm and began to take pictures beneath the lander. There were bright patches, cleared smooth by the exhaust of the retro rockets. And something shining in the sun, an absolutely pure white. Ice? Or perhaps salt? And there was something else that was odd in the images: bumps on the lander legs, bulbous half spheres of some substance that over the course of a few sols seemed to have merged like raindrops on a pane of glass. A team member from the University of Michigan suggested that these were, in fact, water droplets. "Obviously they came from somewhere—they weren't there when we launched," Smith told the press, puzzled. But the temperature had never reached anywhere near warm enough for water to be in a liquid form. It had never even topped minus 25 degrees Celsius.

When the robotic arm began to dig, it hit another white patch, and the rotating rasp at the end of its scoop started throwing out little chunks of bright material. Four sols later, the material had vanished. It had completely vaporized. "It must be ice," declared Smith, for salt wouldn't just disappear. In the next trench, the arm hit a hard patch and simply couldn't dig any further. That patch, also gleaming white, was at the same depth. It was part of a layer. The scientists determined that it was indeed the ice they had come looking for, a pure reservoir, less than thirty centimeters beneath the lander legs.

Next came the analyses, the only time water had ever been directly investigated on Mars. The two main instruments, rising in metallic boxes from the deck of the lander, were called TEGA and MECA. Inside TEGA were eight tiny furnaces, which were designed to blaze forth in the polar cold, revealing the chemical nature of the constituents. When the furnace temperature reached its maximum—far hotter than that of the *Viking* ovens—the vapors driven out by the baking were "sniffed" by the mass spectrometer for tiny quantities of organic molecules. Delivering the samples proved harder than expected, but when the measurements were finally made, TEGA discovered calcium carbonate, the same stuff in Tums. The kind of mineral that was indicative of a watery past.

Phoenix's other key instrument, MECA, had a Wet Chemistry Lab—or WCL, pronounced "wickle." The WCL was designed to dump sugar-cube-sized plugs of soil into a soaking solution, then simmer and stir them. Tiny sensors along the wall of the beaker were fashioned to measure things like salts and acidity. MECA found that the polar terrain was, if anything, slightly alkaline, the kind of soil you'd grow asparagus in. The fact that alkaline soils had been discovered on Mars suggested that the planet might not have been entirely bathed in acid.

The WCL was also designed to assess a range of chemical constituents, including nitrate, one of the most essential for life. A few years before *Phoenix* landed, the presence of nitrate salts, found in extremely dry desert soils here on Earth, had been hypothesized to exist on Mars. Thirty sols into the mission, however, when the WCL did its first experiment, the sensor designed to detect nitrate went absolutely bananas. As it turned out, the detector was also sensitive to another molecule, one that was far rarer. While a small response could have signaled a large concentration of nitrate, a whopping signal could only mean one thing: a massive amount of an obscure salt called perchlorate.

It wasn't a common word in the English language. Smith had to go and look it up. Traces of perchlorate were present on Earth in extremely arid places like the Atacama Desert, measured in grams per hectare. But at the *Phoenix* landing site, just a few handfuls of soil yielded the same amount. *Viking* had detected chlorinated molecules, but the *Viking* science team had assumed they were the residues of cleaning fluids used to disinfect the lander deck. It was now clear that those combusted chlorinated organics were the exact fingerprints that would have been left behind if perchlorate was heated in the presence of organic molecules. The oxygen released would have burned the organic molecules up. The realization had startling implications: Perhaps the *Viking* experiments hadn't been compromised.

I wondered what Wolf Vishniac would have made of it. And Carl Sagan and Vance Oyama, who had both lived through the crushingly disappointing results of *Viking* but had not lived long enough for this. The lack of any simple organics on Mars had been so hard to understand, as organic molecules also have nonbiological origins. At least some organics should have rained down from comets and meteorites. But as the Mars community now realized, perchlorate could have been sitting right alongside them, happily coexisting for billions of years, until they were heated in the *Viking* oven. The perchlorate would have destroyed any trace of organic material, just as it did in the polar soils analyzed by *Phoenix*. We hadn't known it, but we'd been dry-cleaning the samples.

Smith and his team continued to scour the literature. They discovered that perchlorates, as reactive chemicals, might be toxic to future astronauts, but also that perchlorates aren't necessarily bad for microbes. They keep water thawed in a liquid form. Just like the salts we spread on icy roads, concentrated perchlorate salts can act as an antifreeze, massively depressing the freezing point of water to near minus 70 degrees Celsius. Perhaps the bumps on the lander leg were in fact water: water mixed with a bit of perchlorate. Some microbes, the team discovered, could even use perchlorate as a source of energy.

. . .

PHOENIX CONTINUED ITS work through the summer of 2008 and filled its feed with updates. On a day when not much happened: "Feels like scratching nails on a blackboard here." On another: "Just finished hunkering down for a big dust storm." As *Phoenix* wound down, the tweets turned reflective: "I'm not mobile . . . so here I'll stay. My mission will draw to an end soon, and I can't imagine a greater place to be than here." "Whoa, so much sadness about the heater turning off."

Before *Phoenix* died, the instruments recorded falling flurries: water-ice crystals floating gently from high, thin clouds. They were wispy things, like cirrus clouds here on Earth, releasing a sprinkling of tiny particles, a kind of diamond dust. No one ever knew it snowed on Mars, and now it's forever known—a final gift from the spacecraft, a small piece of the permanence of the universe that we don't share in. Most of it vaporized, but a few times, like on sol 109, the streaks touched the ground. There wasn't much of it: "If you melted it all in a pan, [you] would be barely wetting the surface," Smith said. But if you were there on Mars, looking up from Vastitas Borealis, it would have been enough to make the sky sparkle.

Unlike other Mars missions, *Phoenix* was always marching irrevocably toward its end. The light was waning, the temperatures were plummeting. *Phoenix* kept taking measurements and sending lonely updates. But by November of 2008, the weak sunlight could no longer charge the solar panels, forcing the instruments to shut down. *Phoenix* slipped into "Lazarus mode," operating autonomous programs on board but no longer accepting new commands. It was a bittersweet moment. The engineers detected some feeble signals for a few more days, then *Phoenix* succumbed to the polar darkness. The spacecraft's final tweet was: "01010100 01110010 01101001 01110101 01101101 01110000 01101000." Binary for "triumph," followed by a heart emoticon. Within a few months, the air itself froze, entombing *Phoenix* in dry ice.

THERE HAD NEVER been any real hope of *Phoenix* surviving the winter, but the following spring, the *Mars Odyssey* orbiter nevertheless flew over Vastitas Borealis, listening for a beep. No signals were ever heard, even as the lander entered into round-the-clock sunshine that summer. There had only been one part of the mission designed to persist, and it was affixed to the deck of the lander.

It was a mini-DVD containing literature, art, and messages from planetary scientists. Velcro'd to the deck of the lander, the disk was stamped with a paper label, the kind I used to print on a dot matrix when I burned a CD mix for a friend. The collection had originally been assembled for the surface station component of a Russian Mars mission in 1996, but that mission failed, streaking across the sky like a fireball before falling back to the Earth. The idea had come from Louis Friedman, the executive director of the Planetary Society. The top of the disk read, "Messages from Earth. Attention Astronauts: Take this with You." On the left, there was some clip art showing a set of old books, and then at the bottom, text that read, "First Library on Mars."

Included were more than eighty books and stories, poignantly evoking the fact that an entire exploration of Mars had been imagined before anyone left the planet. They chronicled concepts for rocket ships and spacesuits and an astonishing array of other technical details, like solar-powered flight. There were tales about hypothesized contact, telepathy, and utopian existence, and at the same time there were aliens and invasions and bloody wars. Some of it was classic pulp: sword-wielding heroes defending helpless, barely clad women, the kind of science fiction I always found dispiriting, but there were also works from philosophers I loved, like Bertrand Russell.

There was a recording of Carl Sagan, speaking to future Martians from a waterfall near his home in Ithaca. There was Arthur C. Clarke, speaking from Sri Lanka, with a wild cacophony of birds in the background. There were surrealist paintings, comic strips, film posters, and ads featuring Mars in magazines. There were illustrations from novels, maps drawn by Lowell, photographs from orbit, and a plaque from the bridge of the Starship *Enterprise*. There were live recordings of mission control during the *Viking* landings. There was the 1940 recording of H. G. Wells meeting Orson Welles, talking about *War of the Worlds*, the book and the broadcast.

The most recently recorded words were Peter Smith's. In his introduction, he talked about the incredible progress of science—modern medicines, the decoding of the human genome—and he talked about his fears for our planet, his worry that the future might not see peaceful advances in culture, technology, and science. He wondered if our species would survive. He talked about the technology of the future, likely as unrecognizable to us as a laptop computer to Attila the Hun. How peculiar to send a mini-DVD, he mused, which "would not be current even twenty years" into the future. What brave soul might find it? A future astronaut? Might that thin slip of archival silica glass—protected from Earth's rapid decay, preserved on Mars for hundreds of years to come—outlast our entire civilization?

If anything persisted that long, how fitting that it would be our words—our thoughts. One of the oldest pieces included in the collection, and one of my favorites, was Voltaire's "Micromégas." In the short story, written in 1752, a 120,000-foot-tall visitor from another world comes to Earth and initially believes our planet must be devoid of life. But he keeps looking, and when he eventually sees a moving speck in the Baltic Sea, he picks it up with his little finger and puts it on his thumbnail. He discovers it's a whale. He eventually spots another speck, similar in size, and with his magnifying glass, he realizes it's a boat filled with Arctic explorers: ". . . after having commiserated [with] them for being so small, he asked if they had always been in that pitiful condition little better than annihilation, what they found to do on a globe that appeared to belong to whales, if they were happy, if they increased and multiplied, whether they had souls, and a hundred other questions."

I often think about that story, about how much life was teeming beneath the 120,000-foot-tall visitor, about how tenacious he had to be to recognize it. It reminds me of Carl Sagan as he peered at those satellite images, trying to spot roads and fields, using his own magnifying glass of sorts. Do we also have the scale of life completely wrong? In terms of proportion, humans only interact over a few orders of magnitude: micrometers, millimeters, meters, kilometers. How much bigger, or smaller, might life be? And what of time? In Copenhagen, I held cells in my hands that were twenty thousand times my age. Cells that were older than the pyramids, older than writing, older than language. Cells that were alive before the dawn of my species, as the ancestors of *Homo sapiens* trudged the Earth. Cells that persisted as humans walked from Asia to America, as the sea slowly closed. That survived, indifferent, beneath the "Road of Bones," as in the blink of an eye a million people died in Soviet labor camps above. Cells that could remain cocooned until there are no more children, there is no more highway, and there are no more bones.

Then again, what is even half a million years in the life of a planet? Would the giant ever have found life in the Kolyma lowlands? Would he have been an instant too early, or an instant too late? Would he have accidentally stamped out what he was looking for? What kind of magnifying glass would have seen ancient cells slumbering so deep in the permafrost, barely respiring, so easily extinguished?

Like the giant, I'd spent a lot of time trying to peer into worlds too small to see. Once, in one of my teaching labs at Oxford, I'd spent hours observing a tiny fruit-fly larva. Students of developmental biology and genetics, almost without exception, study *Drosophila* as a model organism—the hovering black dots of the adults as well as the eggs, larva, and pupa. The organisms grow quickly, and by playing with their genes, scientists can manipulate them rather mercilessly, causing them to grow eyes on their legs or extra sets of wings and learning a great deal in the process.

In that particular lab, I'd been told to familiarize myself with the larva's functioning under a dissecting scope, then to mount its brain on a glass slide, to study normal cell division, along with its testes, to study germ cell division into gametes. But even after my professor's prompt to begin the dissection, I dallied. I pushed bits of sugar in the larva's confused path and watched as it attempted to climb my tweezers. I couldn't help contemplating the set of circumstances that led this tiny creature to my scalpel. It felt wrong to kill it, but how many lives had I taken inadvertently, without even thinking? There had been countless little genocides that morning alone, just as I walked through the meadow to class. What universes were down there? What noises were too small to hear beneath those blades of grass, where long-horned beetles stampeded like elephants?

I lifted the larva into the pocket of my lab coat and excused myself. I held the slide steady while I walked, glancing down into my pocket every few seconds as I descended to the stairs and found my way to the cafeteria. I left it on a bite of the banana I'd brought for lunch, spat out and dribbled from a napkin, in the base of a rubber plant in the zoology building. I returned to the classroom and gazed for nearly an hour at a blank slide under the microscope, a vast field of emptiness.

WHAT MIGHT I be missing? I contemplated this question in terms of tiny microbes, and the world beyond my doorstep, and the planets I could barely see. Occasionally, my thoughts returned to a little Dutch book from the 1950s—*Cosmic View, the Universe in 40 Jumps*—in which the illustrations expand and contract, traveling outward by orders of magnitude to the edge of the universe, then inward to an atom in the hand of a girl sitting in a chair in the courtyard of her school. There were times, somewhere between the distant galaxies and the atomic microcosms, when I would pause long enough to realize that, like the girl, I was sitting alone.

Then a friend of mine tried to set me up on a date with one of her law school classmates. I initially declined. I was content with my books and my science. I'd never been on a blind date, and I couldn't imagine it being a good idea. "But he knows you," she insisted. So one evening I finally agreed, and a week or so later the guy with the "HELLO" name tag, the philosopher turned lawyer, showed up on the steps to my apartment. He smiled as I cracked the front door and peeked out.

I marveled at his easy confidence. He effortlessly maneuvered his stick shift out of a tight space on the steep hill of my street, then we drove with the windows down to East Cambridge. He'd made a reservation at a restaurant named after Afghanistan's second-longest river, where we shared aushak and showla and a plate of kaddo, baked baby pumpkin seasoned with sugar. At one point, he made me laugh so hard that yogurt nearly flew from my mouth.

How could it last? He was moving to Texas at the end of the summer to clerk with a prominent judge there, to glimpse the front lines of the civil rights issues he'd worked so hard on as a student. I was staying in Massachusetts.

But in the weeks that followed, it was as if a warm candle began to glow. We drove up into the dairylands of Vermont to buy cheese and penny candy. We took the ferry to the Harbor Islands and hiked the green cliffs. I started leaving the lab in the early evenings to go find him wherever he was, which was often playing guitar on a porch somewhere with his law school friends. We'd wander the streets until we happened upon an old bookshop, or bike down to the river to watch the sun set over the tea-colored water.

He was supposed to be studying for the bar exam, but as spring turned to summer, it seemed he was spending all his time with me. My own worry swept into the void created by his calm. I decided I'd better make myself scarce for the final weeks before his test. When a friend invited me on a trek from Norway to Finland, over the top of Sweden, that seemed sufficiently far. In a book she gave me, I read about Arctic creatures and ice, and about how Arctic explorers were always running into one another, despite the oceanic vastness of the landscape. They would wander through uninhabited territory, alone and prepared to remain so for their entire trek. Yet their paths would invariably cross. I was compelled by the idea, how these chance encounters were simultaneously so improbable and so inevitable. It seemed like humanity in a nutshell: that in the moment of our greatest isolation we would be drawn to the same breaks in a landscape, the same dramatic bit of a cliff, and find each other. For ages, I'd wanted to explore that landscape for myself, so I emptied my savings account—the meager remains of my grad school stipend—bought a ticket, and flew to Tromsø. After gathering supplies, I climbed the steps onto a bus east to the place the journey would begin.

As the taiga forest gave way to tundra, I felt like the giant in "Micromégas," thundering about the stands of dwarf willow trees, which were never taller than my knees. The bogs and streams with wild reindeer tracks gave way to windswept heath and lichens as we climbed in elevation. The landscape was littered with boulders, which I kept mistaking for people. The water that trickled all around seemed so pure that I stopped bothering to add iodine before gulping it down.

One morning halfway through the trek, however, my stomach began to tie itself in knots, and by evening, I was vomiting. I felt miserable, but I'd gotten sick before out in the backcountry. The Arctic summer was benign, nothing compared to the Antarctic. The temperatures hovered above freezing. The grass was green. I told myself I was just out for a stroll. All I had to do was get up and walk for a week.

But after a day or two, my body descended into a state of pure refusal. I laid in my tent, barely moving, black flies circling my head. And all I could think about, shaking and sleepless, was this person back in Cambridge: how intensely I wanted to be by his side, how cataclysmic it would be if I couldn't walk my way back to him.

Slowly, through the fervor and fear, I began to realize that despite the vast interiority of any human life, there is a place where our boundaries can dissolve, where the continuity of the self can break, and we can blur into the existence of another human being. He was thousands of kilometers away, yet he was with me: in my dreams as I slept, fevered; up ahead as I walked, gazing out upon the same great horizon. I raced, thirsty, toward the cliffs, tripping and tumbling, grasping for something I couldn't quite name. I was fraying, and he was catching the weave. I felt like those Arctic explorers who had gone off into the vast wilderness only to find themselves suddenly, shockingly, no longer alone.

CHAPTER 10

Sweet Water

ON A WARM November day, the *Curiosity* rover rose up off its launchpad on an Atlas V rocket, billowing fire and smoke, thundering across the lagoons of the Banana River. The fourth in a string of robotic rovers, it was off to Mars to look for evidence of an ancient habitable environment—a benevolent place where life may have taken hold. It was headed to Gale Crater, where, according to orbital data, water in low-lying terrain could have pooled and formed lakes.

The next morning I took a pregnancy test, and the faintest of pluses appeared. I'd hoped to be at NASA's Jet Propulsion Laboratory when the rover touched down on August 5, 2012, but soon after, a doctor told me my due date: also August 5. The timing seemed impossible. Two creatures hurtling through time and space, destined to arrive simultaneously on two different planets.

Despite the tests, the obstetrician's pronouncement, and the fact that an elaborate chemical machinery was kicking into gear inside my body, I couldn't quite believe it. I buckled the seatbelt over my flat stomach and stared at my husband, blinking, as we drove along the

Charles River back to our apartment. The philosopher-turned-lawyer grinned in his boyish way, stretching his arm around my shoulders.

In the days following the launch, the spacecraft settled in for a long cruise. A boost from the rocket's upper stage had pushed it out of Earth's orbit. It was journeying into the deep night, thermally stable and with plenty of power, arcing along its 567-million-kilometer trajectory. Never before had I given much thought to the machinery surrounding the rover—the descent stage and associated hardware, the heat shield, the backshell. It was like a whole second spacecraft, but it had a single purpose, to deliver the rover safely to Mars. Now I too was a vessel. The blood in my veins flowed from my heart to a second heart and back, and soon that heart was beating like a galloping horse. The center of my existence seemed to shift as well, from back behind my eyes—the helm of the starship—to somewhere deep in my abdomen.

I told my family about the pregnancy when they arrived in Boston for Christmas. My parents nearly broke into applause. Emily seemed a little unsure at first, but, sensing the excitement, she bounded in for a hug. She soon found a pregnancy book from the Mayo Clinic next to my bed and studied it intently for days, amazed that such a book existed. In the margins of the pages, she suggested names, including several that were an homage to her favorite actor, Tim Allen. Whenever she saw something that seemed important, she brought the book over to my husband—"Brother," she called him. "You are her partner and her coach," she told him, thrusting an image of a breech baby, dangling one foot into the world, onto his lap.

As the book instructed, I made the suitable course maneuvers. I stopped eating soft cheese and working with chemicals. I avoided caffeine, secondhand smoke, X-rays, kitty litter, hot tubs, and alcohol. I took a prenatal vitamin every morning and a DHA supplement. I shortened the one trip to the field I had planned, and when I arrived at the airport, I let the TSA pat down my body. I wore sensible shoes and watched my step. Instead of running in the mornings, I started to walk, and then began to walk slowly.

My restless little one began to somersault through the weeks, up until the night before he was born. He was in an "unstable lie" when I reached term, flipping every couple of days, and the doctors began to worry about the umbilical cord being compressed. There would be surgery, scheduled for the first available slot. I'd been told that the day an otherwise-healthy woman gives birth is the most dangerous day of her life. In all my years, I had never broken a bone or needed an X-ray. I hadn't spent any time within the walls of a hospital.

When the time came, I shuffled to the operating table and sat on the edge as a resident tapped a needle between my vertebrae, then pushed anesthesia into my spine. I lay down beneath the bright lights. My husband and mother took their place at my side, and everything piled together: the sleepless night, the drive to the hospital, the difficulty with the IV. Then the scalpel dipped into my belly. There was the sensation of pressure but not pain, and only a moment later, the whoosh, the fall of my left hip back to the table, the tinny sound of a cry clamoring from my baby's small lungs.

As my sister foretold, he was born breech: feetfirst into the world. He was pink and trembling, with beautiful navy eyes. Hours later, unable to walk, mooning from the painkillers, I clutched him tightly, trying to shoo the nurses away. I did not want to rest. I did not want to go to breastfeeding class. I wanted to hold him. His body had just been clipped free of mine. I explained that I'd had a C-section, fully expecting that to be a point in my favor as they tried to take him to the nursery. Of course he had to stay in my arms, having been pulled so quickly into the light.

I'D BARELY GOTTEN home to our apartment when, in the dead of night, *Curiosity* landed. Warm summer air was streaming through the fan in the window, and everyone had fallen asleep—my husband, my mother, my tiny son. I cradled him in my arms as I watched the live stream on the screen of my laptop, illuminating the room with a faint blue hue. We were perched on a big chair by the bookcase. His taxed little body, starving for milk that hadn't come in, had collapsed against me.

Wide-eyed, I caught glimpses of friends and mentors among the rows of matching periwinkle polo shirts at JPL. They had been eating peanuts for luck, a long-standing tradition. *Curiosity* was landing via a "Sky Crane," a sort of technological stork. Unlike *Spirit* and *Opportunity*, which bounced to a stop enveloped in airbags, the sheer mass of *Curiosity* required a soft landing. It was the size of a Mini Cooper, a metric ton, a car falling from space—nothing like it had ever been attempted.

The minutes ticked by as the spacecraft barreled toward the surface, nested inside that descent stage with its four steerable engines. The whole landing was only seven minutes, about the same time it took the obstetrician to tug my son from the womb. Shortly before 1:30 A.M. Eastern time, the parachute flew open. The rover dropped to the ground with a bridle and umbilical cord, both of which were promptly severed when the wheels touched the red dusty terrain. After the rover separated safely, the rest of the bundle powered away at full throttle for a crash landing off in the distance.

Once touchdown was confirmed—the signal that *Curiosity* was indeed on the surface and everything had gone according to plan—an all-male panel of scientists took the stage at von Kármán. I slowly rose from the couch and limped off to bed.

As I passed the bathroom, I caught a glimpse of myself in the mirror. My body was ripped and bruised and bandaged, and I could barely recognize its swollen contours. But in my arms, inside a blanket covered with caterpillars, was an immaculate baby. I swaddled him in the darkness of the bedroom. When I fumbled the final tuck, he let out a small cry. My husband stirred and quietly pulled back the covers for me. Then everything was silent again, except for the sound of our son's tiny breaths. As I lifted him into his bassinet, I felt the kernels of his backbone. I touched his fingers, which curled like the tendrils of fiddlehead ferns.

CURIOSITY TOUCHED DOWN in an unknown land. Gale Crater was over 150 kilometers across, but the distant walls of the chasm cut the horizon, making everything feel a little closer. Not far from the landing site were dark dunes, and behind them, a breathtaking mountain, soaring higher than Mount Rainier above Seattle. Gale Crater was chosen because of the towering stack of sediments in its center. Mount Sharp was five and a half kilometers high, and thought to be one of the thickest geologic records in the entire solar system. The stratigraphy recorded a breathtaking stretch of Mars's climatic and environmental history, including ancient terrain filled with soft clays.

It was a perfect place to explore. For years leading up to the mission, NASA's mantra for Mars exploration had been to "follow the water," but we'd found the water: ancient river channels, hydrated minerals, gullies, near-surface ice. With *Curiosity,* NASA was taking a new step, moving beyond the search for water to look for other signs of habitability. *Curiosity* was trying to understand the context: Was there a chance that microbes might have survived here? Were the right ingredients in place? If so, for how long? Long enough for life to gain a foothold? Were there simple organics, the bricks from which a house of life could be built?

The rover came to rest just north of Mount Sharp, facing east-southeast. The ground was the color of an apricot, stippled with small pebbles. Like my surgical team, the engineers had done a fabulous job—in their case, setting the rover down just two and a half kilometers from the center of the landing ellipse. It would be a long drive to Mount Sharp, which was still a few kilometers away, and *Curiosity* couldn't set off immediately. Everything had to be tried out and confirmed. The rover's computer software had to be updated. Then the communications link had to be tested. The team heard the first human voice beamed up to another planet and back, a message of congratulations from NASA administrator Charlie Bolden. The wheels got to practice driving forward, then back. The rover's arm stretched out to test its components—the drill, the brush, the sieve. The instruments were checked to make sure they had made the journey safely. It seemed that every little thing sent the rover into SAFE mode, setting off a cascading series of errors that would leave it frozen, unable to move. The team would momentarily panic, then realize a particular range restriction or temperature limit didn't need to be as stringent, at which point they would write an override function into the code.

. . .

BACK IN BOSTON, I too was getting into the swing of things, becoming less terrified that I'd somehow drop my son, let him fall from the changing table, forget him in the car seat, or overheat his little body in blankets. I learned that a particular pitch of crying meant he wanted to be held and how a hand to an eye was a clear sign of sleepiness. My own catastrophic exhaustion abated, slowly, replaced by the wonder of his circus tricks. He'd splash the bathwater with his hands open like stars, and sometimes in his sleep it would look like he was directing traffic. Often he would just gaze at me, his face so open and innocent, and all I could do was look back. Giving rise to a new human life felt like the ultimate experiment. What happens when you combine half your DNA with half the DNA of the person you love most? Who results? Who does he become? How does consciousness spring from within the plates of a small soft skull?

For most of that first year, I carried him everywhere I went—to my office, to the library, even to a conference or two. We had to be together every three hours anyway. I'd bundle him up and lift him into the baby carrier. He'd nestle his head into the warm space beneath my chin, and we'd head out into the day. While *Curiosity* was making its way to the base of Mount Sharp, covering kilometers of open space, we were pacing the inhabited eggshell of the Earth together. I had spent most of my life inside my head, and most days it was a welcome change to simply be in the moment, holding his tiny hands.

I still followed the mission from afar, momentarily trading my immediate world for the depths of space. When I did, I sometimes felt a pang of sadness that Mars might be slipping away. Opportunities only came around so often. The planets aligned and then swung back apart. They waited for no one, and NASA's next rover wouldn't land for another eight years. I was only a postdoc, and not a terribly productive one. Even if I scrambled, it wasn't obvious to me that I'd catch up.

I FELT THAT longing most acutely as *Curiosity* began making major discoveries. The rover happened upon the rounded shape of river rocks, then a broken sidewalk of outcrop, fragments of cemented bedrock encasing stones and sand that had once tumbled downstream. It was an ancient riverbed, winding down from the northern edge of Gale Crater. It had once been rushing with water, perhaps as deep as my hip.

The rover continued on toward the junction of three different types of terrain that came together in a low-lying depression about four hundred meters away from the landing site on the way to Mount Sharp. It was named Yellowknife Bay, after Yellowknife, Canada, a place I'd always wanted to visit. Both Yellowknife, Earth, and Yellowknife, Mars, sat upon rocks that were four billion years old.

Where the ancient rivers emptied to standing water, the rocks took on a much different character. As the momentum of the rushing water gave way, fine-grained particles had a chance to slowly settle. The rock unit that *Curiosity* discovered, the Sheepbed mudstone, was the bottom of an expansive lake, with mud that would have squished in our toes. At one point in Mars's ancient past, the lake must have swelled and evaporated—filling, emptying. The crater's central peak might have even stood as an island.

IT WAS IN 2016, three and a half difficult, searching years after *Curiosity* landed, that I finally had the chance to join the science team. In that time, my son had started preschool and his younger sister had been born. And somehow, in the midst of it, I'd moved to Washington, D.C., and started a job as an assistant professor of planetary science.

NASA's Goddard Space Flight Center, which served as the base of operations for one of *Curiosity*'s main instruments, was only twenty-five kilometers northeast of my fledgling lab at Georgetown University. I knew Maria had begun her career there, and I instantly realized why she liked it so much. There were so many talented people and an impressive sense of collegiality. There was "Breakfast and Learn" on Tuesday mornings and "Colla-BEER-ation Hour" on Friday afternoons. I started spending lots of time there, and within a couple of months I was invited to become a visiting scientist with the Planetary Environments Lab.

The lab chief, a brilliant chemist named Paul Mahaffy, had built the Sample Analysis at Mars instrument, also known as SAM, with help from French scientists. Paul had grown up in Eritrea, the son of American missionaries. When he wasn't studying, he joined his six siblings in eating injera, collecting scorpions, and watching the laughing hyenas quarrel with local dogs. His village was near a huge granite spire, Emba Matara. At its peak was a great iron cross, as tall as a building. As a child, he would scramble to the summit, then climb to the top of the cross and sit there listening to the wind.

He now ran SAM, which felt like the beating heart of the mission to someone interested in the search for life. It was one of the most sophisticated spacecraft instruments ever made, weighing nearly as much as all of *Curiosity*'s other instruments combined. The whole rover chassis had been designed around its gold-plated box.

One of SAM's jobs was to measure isotopes, different versions of the same atoms whose ratios helped reveal how much air and water were lost to space. By bouncing a laser back and forth, SAM could scour puffs of Martian air for methane, a gas that is made almost entirely by microbes here on Earth. Methane has bedeviled Mars scientists for nearly half a century. *Mariner 7* scientists announced the discovery of a methane plume near the South Pole, only to retract their results a month later. In the early 2000s, powerful telescopes in Chile and Hawaii reported detections. *Mars Express,* an extremely successful European orbiter, also picked up traces, though an order of magnitude less intense. Then, inexplicably, the methane vanished, and for years no methane was seen at all.

SAM then found it again, uncovering not only its presence but a seasonal pattern, with levels rising dramatically in the summer. There were several possibilities. It could be coming from a dance of water and rock deep in the subsurface, a purely geological process. It could be ancient, created long ago and locked away in matrices of melting ice. Or it could be the exhalations of a small, still-active biosphere.

SAM also carried a gas chromatograph-mass spectrometer, a core tool for understanding chemistry, particularly organic chemistry. I'd known about the machines since I was a kid. My father had earned an undergraduate degree in chemistry at the foothills of the Cumberland Mountains, at Berea College. As a technician in a medical examiner's laboratory for the state in Frankfort, Kentucky's capital, he ran and repaired the gas chromatograph-mass spectrometers that helped pinpoint chemicals in dead bodies during autopsies. I remember going to work with him one day, not long after the state decided to exhume President Zachary Taylor to test the theory that his sudden death in office could have been due to arsenic poisoning. I stood in my Keds in the tiled space in front of a laboratory freezer as my father opened the door. He passed me a 141-year-old toenail. I held the vial above my face, turning it in the light. I heard the tissue clink against the clear plastic. I was eleven years old, and I was holding part of a president—one who wasn't poisoned, as my father's lab discovered.

Although I could sense the enormous power of these instruments, I had no inkling then of their potential as tools for space science. Years later in graduate school, when the opportunity to work on a mass spectrometer came along, I jumped at the chance. I began to plow through papers about how scientists were using them to uncover traces of life in ancient rocks. Under the right circumstances, certain molecules, like lipids in cell membranes, could stick around for billions of years. Even if a group of atoms here or there fell off, like fingers or toes, the backbone of the molecule could still tell stories about the cell it came from, the same way we can learn about dinosaurs from dinosaur bones. Patterns among very simple molecules could also serve as strong indicators of life. Even without patterns, the mere detection of organics was key to determining the conditions for life as we know it.

When the *Viking* landers carried the first mass spectrometers to the surface of Mars in the 1970s, they found no definitive evidence of organics. But *Curiosity* had three things going for it: It had a far more sophisticated detector in SAM, with a resolution of better than one part per billion. *Curiosity*'s landing site was specially chosen for the fine-grained clays that could have once trapped and preserved organics. Also, the rover could sidle right up to the best sampling locations and use its drill to access the protected interior of rocks.

When *Curiosity* began carving into the mudstones, it kicked out not hard orange oxidized rock but instead a soft gray powder. A pinch of pulverized clay, about half the size of a baby aspirin, wound its way down a sieve into the belly of the rover. There, diagnostic analyses revealed that the rocks were made of clay minerals that typically form under neutral pH conditions. Mars hadn't been thoroughly soaked in acid after all. And unlike at Meridiani, there was hardly any salt in these mudstones. All six of the elements required for life as we know it were present in the sample: carbon, hydrogen, oxygen, nitrogen, phosphorus, and sulfur. It was the context we needed. Not only was there water, it was the right kind of water, in the right kind of place. The Sheepbed mudstone was just where we should be looking for life. At the press conference that followed, John Grotzinger, the mission's lead, announced that a habitable world had finally, definitely been discovered. "If this water was around and you had been on the planet," he said, underscoring his point, "you would have been able to drink it."

What's more, there was finally an indisputable detection of organic molecules. SAM's tiny oven heated up the sample, turning the molecules in the mudstones to gas, then whiffed them into a long thin tube about the width of a dust mite. One by one, they popped out the other end, where the mass spectrometer identified them: simple compounds, chlorinated from interactions with perchlorate from the surface. And discovered later, even more-complex molecules, bound together by sulfur. The high levels of the detections, up to three hundred parts per billion, set to rest one of the longest-standing mysteries about Mars: The building blocks of life were indeed there.

BY THE TIME I joined the science team, *Curiosity* had holes in its wheels, and its poor drill was beginning to show signs of wear. It had been roving for four years. But the mission was about to enter into one of its most interesting phases: an ascent of the Vera Rubin Ridge, named after the extraordinary scientist who discovered the first evidence of dark matter, mysterious particles thought to comprise roughly 85 percent of the total matter in the universe. As a girl, Rubin had learned to tell time by how the stars outside her window wheeled across the sky. As a young woman, she was encouraged to pursue a career not in astronomy but in painting astronomical objects. Undeterred, she went on to earn a PhD at Georgetown and stayed on as an assistant professor at a time when many astronomy departments were not open to women. She was averse to sharp elbows, gravitating to a research question that "nobody would bother [her] about." Working in this self-imposed obscurity, she transformed cosmology. She proved that the vast majority of the universe is invisible, a finding I have always found poignant.

Near the end of her life, she wrote that she had succeeded in her two goals: "to have a family and to be an astronomer." She had four children, three sons and a daughter. The eldest, born when she was just twenty-two, even became a planetary geologist. He now works on *Curiosity*'s science team, striving to understand the formation of the ridge named for his remarkable mother.

Rubin's old observatory is still on campus, on a hill just opposite my lab. I pass it sometimes in the mornings, as one of my children's favorite things is to see the pollywogs in the adjacent pond before preschool. On the way back down the hill, they stop to pick up pine cones for their teachers and to stuff pebbles into their pockets. They are full of questions, shining their attention like lamps on everything before them. They spend all their time, every moment, stitching together clues about how things work. Watching them, I'm reminded how we're born knowing so little and understanding even less about the context of our world. They look to me for explanations, for words they don't know, for help connecting the data points. All of their questions seem to contain another "why." Why is it this? Why is it this and not that? It reminds me of something Rubin once said: "I'm sorry I know so little. I'm sorry we all know so little. But that's kind of the fun, isn't it?"

I want time to stop in these moments, with their tiny fingers laced into mine. Even though we walk slowly, pausing constantly to inspect things, we always arrive, and at the sight of their small friends, they run from my arms. I drop their lunches in their cubbies, then linger, watching them for a few more moments through the windows of the building, my fingertips against the panes of the glass.

But eventually I walk away, and before I know it, I find myself on Mars, where *Curiosity* is still roving. From time to time, I get to help guide the rover. On those days, I either drive out to Goddard's SPOCC—the Science and Planetary Operations Control Center, where SAM is operated, where life-sized cutouts of Spock from *Star Trek* decorate the hall—or I remotely access JPL's server from my office at Georgetown. Downlinked images, caught by the giant dishes of the Deep Space Network, stream into the viewing software. They too transport me. They take my breath away. As the team decides what measurements to make and the engineers check the commands, I wonder how the rover is doing all those millions of kilometers away. What would it be like to be standing there, to hear the sounds of the instruments warbling in the darkness?

When those days end, when eventually I realize I must go, again it feels almost impossible to leave, one love tearing me from another. Sometimes I'll pan slowly through the last mosaic image of the Martian landscape, stopping at a certain angle where the land meets the sky. This is what I'll see when I next flip open my laptop, my teleporter. What comes next? What's the best way up the mountain? Is this new or something seen before? What does it mean? I carry the image in my mind, a wilderness stretching off into the horizon, vast and full of possibility.

CHAPTER 11

Form from a Formless Thing

DEEP IN THE forests of western Bosnia is a village named Jezero. In many Slavic languages, "jezero" is the word for lake, and jezeros stretch all along the Adriatic. They fill the Julian Alps, dotting the spaces between low-lying meadows and frog-filled caves. There are sinkhole jezeros and karst jezeros, glacial jezeros and jezeros linked by a hundred waterfalls. But the jezero in the village of Jezero is green and quiet, almost mythic. It's preternaturally still, rumored to be heavy with deuterium. The surface reflects the clouds like a piece of polished glass.

Small craters on Mars are named after small towns, and on the western edge of Isidis Planitia is a small crater named for this small Bosnian town. Early in Mars's history, Jezero Crater also held a lake with water that reflected the sky. Two rushing rivers emptied into the cavity—one from the west and one from the north. The lake was deep, its crater floor plunging hundreds of meters down from the rim, which on one unexpected day billions of years ago suddenly fissured, unleashing a catastrophic torrent of water over the side.

Among the bevy of spacecraft that will soon launch toward Mars is a NASA rover that will land on the spill of lava that covers the floor of Jezero Crater. The mission's breathtaking goal is to collect samples from Mars to be brought back to Earth at some later date, samples that not only may harbor signs of ancient life but also could give us an unprecedented look into the history of our solar system.

Here on Earth, our record of deep time has been forever lost. The seas have lifted into rain, and the rain has beaten the surface bare. Our planet has swallowed itself, plate by plate. Our original crust has almost completely disappeared; all but a few patches have been dragged back into the interior. The small blocks of rock that remain—in the cherts of Australia and the greenstone belts of Greenland—have been cooked, mostly beyond recognition. Our early days are irrecoverable.

Mars, however, is all past. It is as if time stands still. There are no plate tectonics, no large-scale recycling of rocks. The rivers have stopped; the temperatures have plummeted. On the scale of humanity, Mars has been constant. To be sure, there is weather, like the spectacular dust storms that come and go. Barchans of sand shift across the surface. The polar caps wax and wane. The planet's spin axis arcs into a deep bow every hundred thousand years or so. Yet the land beneath remains.

The place we're now visiting with spacecraft is almost the same world that it was three billion years ago. As a result, the right samples may even help us fill in the gaps in our own planet's history. Right now it's unclear what prebiotic chemistry dominated the early days of the rocky planets, or what dance of reactions made the first protocells. Perhaps life sprang from geothermal fields, with repeated cycles of wetting and drying helping to form complex mixtures of important molecules. Or perhaps not. The samples the rover collects might hold within them the echoes of the beginning of life, entombed deep in the planet's ancient rocks.

The rover's chassis is the same as *Curiosity*'s, but it will carry a different scientific suite, even a small helicopter to test the viability of airborne craft. The rover's two-meter-long arm, laden with new coring tools and instruments, looks like an outstretched lawnmower, and is just as heavy. Over at least two years of operations, the turret will drill several samples of rock and place them carefully in sample tubes, each about the size of a penlight. The rover will then deposit the tubes in a little pile on the surface. The cache will remain there for many years, glinting in the sun, until a fetch rover comes and launches them into orbit—to be caught by a passing spacecraft and brought home.

Like the rocks we carried back from the moon, Mars rocks will be analyzed for decades to come. Once we have them in hand, we'll have them forever. It's been nearly fifty years since the last humans walked on the moon, yet the Apollo samples have been examined again and again, particularly as new tools and technologies have been developed. In that time, we've discovered astonishing, unexpected things, like the precise age of the moon and the fact that the rocks carry an indelible record of the history of solar activity.

If we are going to archive samples of Mars, now is the time. One day, potentially one day soon, there will not only be rovers and robots, there will be people exploring the planet. SpaceX is already calling for a million passengers, sent on a thousand spaceships. But unlike rovers, which we can bake and clean, humans will shed life left and right, sloughing off cells, littering the planet with biological material. The next decades are thus critically important for the search for life because the window to explore an untrammeled planet—a pristine record of the past—is closing.

JEZERO CRATER IS home to a relict river delta, and that is the reason it was chosen as the landing site. There are two deltas, in fact—but the larger and more magnificent, which fans gorgeously out to the east, accumulated rocks and debris along the crater's western rim.

In many ways, deltas are the perfect place to investigate life. As rivers move into steady water, they slow down and spread out. Frictional drag is what suspends small grains of sediment and keeps them leaping along. But when the water slows, the somersaulting particles fall. The grains sort into sizes—coarse sand settles first, then silt, then clay. The finest-grained material is the last to settle and the most likely to trap things. These gooey clays bind and bury organics.

They harden into impervious mudstones, and the molecules within them are protected from oxidation and other forms of chemical attack.

This is what we hope we'll find at Jezero. The large delta there was fed by headwaters that stretched for scores of kilometers, all the way to the horizon. The clay-bearing distal edge of the delta is one of the rover's main targets, its rich bottomset beds offering a chance to find the archived traces of ancient life.

Herodotus named deltas after noticing that the triangle shape at the mouth of the Nile resembled the Greek letter delta, Δ. He was a great historian, determined to prevent the traces of human "events being erased from time," but he was also an explorer. Reaching for the edge of his ancient world, he first sailed to Egypt—the "gift of the river"—in around 450 B.C. He documented the trip in an "autopsy," or personal reflection. One of the first things he noticed was a great "silting forward of the land" as far distant as a day's sail from the shore. He described how the river spilled out plaits of fine-grained clay when it emptied into the Mediterranean, how you could let down a sounding line and bring up nothing but mud.

The same sludge carpeted the delta, transforming the desolate Sahara into a place with flamingos. The fine-grained material was filled with nutrients, perfect for growing crops: durum wheat, emmer, flax, barley, rape, black mustard. From the ground rose chicory and parsnip and the spices of caraway, anise, and hop. From the end of the Paleolithic, the ancients plowed and seeded during the long winter growing season. In the spring was the rich harvest, when flint-bladed tools would reap the land.

Then in summer, the fields would flood. The Egyptians had a word for it, known today as *akhet*. (Like an eroded relic, the word's vowels are lost to history, with only consonants remaining in the hieroglyph.) *Akhet* was the inundation. Under the dog star, Sirius, with the Nile swollen, people mended their tools and tended their livestock. They lifted the mud from beneath the water to make pots. Among the sycamores and reeds, they formed the wet clay with a kind of potter's wheel, hand turned. They smoothed the surface and fired the receptacles in makeshift kilns. They learned how smoke could darken the surface and how oxides of copper could brighten it. They decorated the jars and jugs with pictures, thoughts, and poems, then filled them with water and wine, oil and grain. They carried those vessels with them, often into the grave, where the patterns would remain thousands of years later, their colors still resplendent.

It was the Pelusiac branch of the Nile, off toward distant Sinai, that Herodotus took as he voyaged up the delta. He sailed alongside primeval thickets of papyrus, spangled with feathery umbels. Stalks sprang from the shallow water, some reaching almost as high as five meters. The fens had a special place in Egyptian cosmology. The world was created when the first god stood on the first piece of land: a rise that appeared, like the end of *akhet,* from the boundless dark water.

The fens were a dark and mysterious place, and from them sprang the germs of creation. Like the black mud, they were a "gift of the river," particularly their papyrus stalks. He noted how the Egyptians ate the young shoots, roasted in red-hot ovens. They made garlands from the flowering heads and mashed the reeds into the seams of boats. They pressed out sails from the spongy white pith, affixed to masts of acacia. And, importantly, they made paper, the perfect surface for recording language.

Herodotus's writings found their home on papyrus scrolls in the Great Library of Alexandria, on the other end of the delta, where winter waves and longshore drift flushed silt eastward. The library served for generations as a hub of unbroken scholarship. A great lighthouse signaled visitors in the night, reflecting fire with polished-metal mirrors. When ships arrived into the port, manuscripts were sent to be copied by scribes. In time, the library's collection grew to tens then hundreds of thousands of scrolls.

Never before had such a powerful repository of knowledge been amassed. A gathering of ideas, like the gathering of sediment. A place of sifting, sorting, synthesis. Traditions came in from the Persians, the Babylonians, the Assyrians, and the Phoenicians. And in that fertile space, new masterpieces emerged.

Among them was one of my favorite books, *Euclid's Elements,* spare and uncluttered. Euclid didn't invent the mathematics underlying the *Elements,* at least not all of it, but he synthesized the work of his predecessors in a novel way. There at the edge of a delta, he laid out thirteen sections of geometry and arithmetic: definitions, postulates, theorems, proofs. He charted a course through plane geometry and incommensurability, from the infinitude of primes to the cubature of pyramids, cones, cylinders, and spheres. It was an internally coherent system of mathematics, built from first principles. It was an unprecedented accounting of the physical universe.

Like most American students, I learned about Euclidian geometry in school, but those lectures and assignments didn't begin to capture its majesty. The summer I was thirteen, just a few weeks after I'd returned my geometry book to my teacher at Morton Middle School, I drove with my family to eastern Tennessee. My eighty-year-old grandmother was in decline, and my mother was summoned to help. My father couldn't miss work and needed the car, so he ferried my mother, sister, and me the four hours up and over Jellico Mountain.

After we pulled into the yard, I opened the car door to the faint smell of submerged vegetation, the slow-moving Tennessee River just half a kilometer away. As I walked to the porch of the small clapboard house, bits of the old weathered smokehouse got stuck in my jellies. The air was thick and hot, the fields overgrown. The coal trains were the only entertainment. When I'd hear one coming, I'd leap up and burst through the screen door, just as my mother and her seven siblings had as children. But aside from when the engines shook the house into a whirl of noise and wind, the summer was shiftless. As the days turned to weeks, my mother realized I needed something to distract me.

"You like math, don't you?" she asked one morning as I paced the narrow intersection of the house's three rooms. She called back to Kentucky and arranged for materials to be mailed to us. Our state university had put together a teach-yourself-by-mail course for its students, designed for freshmen entering with a weak foundation in high school subjects. A few days later, a set of small blue books arrived. "This might be fun," she said. "Work through the problem sets, then we can walk to the post office and mail them back to Lexington."

So that summer, cross-legged on a horsehair couch, I taught myself algebra II and trigonometry. My grandmother, with only a grade school education, looked wide-eyed as I drew and erased conic sections. It hadn't been my mother's intention, but the summer of math meant I entered high school ahead of my peers that fall, having already learned arithmetic with polynomials, complex numbers and logarithms, and trigonometric functions. It meant I finished high school math early.

The fall after I got my driver's license, I started driving over to the University of Kentucky. I began spending time in the math department, and it was there that I met a professor named Dr. Brennan. He had short white hair and wore his shirts tucked in, his pants high. He had a kind smile, and I'd often see him walking to the pool to swim laps, humming contentedly. He agreed to let me take his number theory course. It was my first foray into pure math, and with it, the world opened.

The course was essentially *Euclid's Elements,* starting with Book 7. But there were no numbers. There was nothing to memorize. There were just words: the posing of problems that at first seemed elementary, easily understood—for example, *prove there are infinitely many prime numbers*—but the more you thought about them, the more complex they became. Being inside a proof was like being inside a great and turbulent ocean. I would bob about—adrift and confused, struggling to stay afloat. Then there would be a moment when everything would still, the waves would flatten, and I'd see my way to shore.

I started meeting Dr. Brennan for coffee at seven or eight in the morning, before class began, and we'd work out proofs on the backs of napkins—congruence, divisibility, the building of integers, the construction of perfect numbers. I'd never felt anything more austere and beautiful than when pages of equations would collapse into QED—*quod erat demonstrandum,* the very thing required to be shown. It was a Kantian ideal, a wondrous kind of knowledge that both existed in the world and was also independent of anybody's particular experience in that world. It was a complete, self-contained system based not on faith but on reason, and just like so many students before me, I only needed to be shown its inner logic to know it was true.

Around the turn of the nineteenth century, a mathematician

named Carl Friedrich Gauss set out to perfect what he saw as the only potential blemish in *The Elements:* the fifth of Euclid's five starting postulates, the fifth of the five things he assumed to be true at the outset and proceeded to use as the basis of his reasoning. The first four were plain as day, like the fact that a straight line could be drawn between any two points. But the fifth, the parallel postulate, was a bit more complicated, a little less self-evident than the others. It hadn't fully satisfied Euclid either. He'd avoided using it as long as possible, proving the first twenty-eight propositions in *Elements* without it. When Gauss started tinkering with the consequences of a geometry where Euclid's fifth postulate didn't hold, where more than one line could be drawn through a given point parallel to a given line, the results were vexing. He was searching for a contradiction, some sort of logical proof it was impossible, but try as he may, a contradiction never fell out. In fact, he had defined a new type of geometry, and slowly it dawned on him that it was every bit as valid as Euclid's. He realized that Euclid's geometry, which seemed so intuitively right, might not hold. Gauss, never one for controversy, kept his doubts a secret until years later, when others slowly came to the same conclusion.

Neither Gauss nor the geometers who immediately followed, who defined not one but multiple non-Euclidean geometries, lived to see the twentieth century, when the theory of general relativity was published. In it, Einstein posited that stars and planets dimpled the fabric of the cosmos, that space-time was curved and bent by matter and energy. Subsequent deep-space experiments proved him right and, in so doing, showed that Euclid's fifth postulate, the parallel postulate, didn't actually describe the physical world. For all its beauty, for all its explanatory power, *Elements* wasn't real.

I THINK ABOUT this often when I think about the search for life—what we know and what we can trust, what we believe and why we believe it. Deltas are places where we know life to be drawn, sorted, and preserved, where evolution is recorded. We know how to handle a delta. We know where the clays spill out and where to find the bottomset beds. We hope that, like on Earth, they harbor concentrated biological material—quickly deposited and quickly entombed. And Jezero is home to ancient sedimentary rocks that are far better preserved than the most ancient sedimentary rocks on Earth. In all likelihood, the rocks were abandoned quietly and have sat quietly ever since. And yet, as great as this delta is, there are so many other parts of Mars to explore, some just beyond Jezero's rim.

Virtually everything we know about the history of life on Earth we know from the sedimentary rock record—from sedimentary stratigraphy—but that's largely because of the success of photosynthesis, which gave rise to a tremendous proliferation of biological material. As some of it was buried, it imprinted itself like a tattoo onto the skin of our planet. Yet photosynthesis evolved late, well over a billion years into our planet's history. Before it began on Earth, the simplest, earliest forms of life survived not off the sun but off chemical sources of energy. What if photosynthesis never got going on Mars, or didn't last? It's possible that Mars never had a surface biosphere, for by the time photosynthesis evolved here on Earth, Mars was already a pretty hostile place. The planet was bathed in radiation, and the temperatures were frigid, save for the momentary warmth of meteorite impacts. If there wasn't an extensive surface biosphere harnessing the energy of the sun, then we're not really sure what we'll find in a surface delta.

There are, however, possibilities other than photosynthesis. If life on Mars had a chemical source of energy in the lake, it might have been able to withstand the unforgiving environment. And if it couldn't, it may well have retreated underground. Here on Earth, the vast majority of microbes are located in the subsurface, stretching their tendrils deep into our planet. We've pulled up strange microbial life from some of the deepest subterranean mines in the world, and there are more of those little homes on Mars than on Earth. The rocks are more porous, less compacted from gravity. The science is still new, but we're starting to understand the kind of life that lives in dark rock, and it's entirely possible that such life might have existed on Mars.

It is for this reason that, after completing the primary mission at the delta—a full two years of exploration—the rover will head into the unknown. At least a dozen and a half samples will have been cached, a pile of tubes left on the ground back on the floor of Jezero. The rover will then carry a set of spares off into its new hunting grounds.

In the distance, some nine hundred sols away, is the edge of Northeast Syrtis, a site called Midway. The landscape at Midway is a little like a Dalí painting. There, giant hunks of terrain were thrust right out of the ground by the Isidis impact. These megabreccia are colossal and beautiful blocks of primordial crust, records of the very earliest days on Mars.

There are ridges and mesas, hundreds of meters in size, that tower over polygonally fractured terrain. Tucked inside are clays and carbonates but of a different form. The minerals there are most likely to record evidence of subsurface life. We'll peer into fractures and veins and look for physical interfaces as well as chemical gradients. We'll search for evidence of exhumed subsurface aquifers. We'll scour deposits of serpentine, a mineral that forms from hydrothermal activity down in the depths of rocky planets. Maybe there we'll discover an underground mausoleum, some single-celled version of the catacombs beneath Paris.

To me, that's the most exciting part of any mission, venturing into the unknown. There are many risks, of course. The rover might get stuck or caught in a dust storm or simply break. It's hard to plan for the caching of samples from regions about which we know so little. But if we make it that far, sample return could be an unprecedented catalyst to our effort to understand "life as we know it."

Even more intriguing, for me, would be finding life based on new biochemistry. We know how to look for particular classes of molecules, search for recognizable patterns, but those molecules might be different on Mars, and those patterns might not hold. We are still struggling to contend with the truly alien, to recognize and interpret signs of "life as we don't know it." But we are making progress. We're learning how to search for evidence of things like chemical complexity, unexpected accumulations of elements, and signs of energy being transferred, things that could shine a light on life even if it was very different from our own, even if it was built on an entirely different molecular foundation. It's one of our biggest intellectual and practical challenges—like trying to imagine a color we've never seen.

IF I'D LIVED and worked in ancient Alexandria, the idea of searching for life elsewhere probably wouldn't have even occurred to me. Before Euclid was Aristotle, Alexander's powerful tutor, whose beliefs, like the *Elements,* hung like a spell over the Western World for nearly two millennia. Aristotle rejected the idea that there was a beginning of the universe or a beginning of time. He rejected the idea of the atom, a thing "uncuttable," and the atomists who advanced it. The world was not particles and forces. It was not built from inanimate parts. Everything had a fundamental purpose. Objects with fire in them were drawn upward because of their essential nature, just as objects with earth in them were drawn toward the ground. There were some quiescent bits, like rocks, but they too had an essential nature in Aristotle's view, perhaps one we just hadn't yet discovered. The heavens—everything above the moon—were pure and perfect. They were a separate realm, made of "quintessence," and they were wholly apart from human experience.

Aristotle's ideas dominated Western thought for centuries, right up until the beginning of the Enlightenment. But, like Euclid's, many of them were wrong. The Earth wasn't still, even though no one could feel it moving. Objects of different weights didn't reach the ground at different times. Flies couldn't spontaneously generate from meat, nor could eels from mud. The blood of men was not, in fact, hotter than the blood of women, and men didn't have more teeth.

The scientific revolution ushered in a penchant for investigation. There were new ideas and new tools. No longer was knowledge a product of the human mind looking inward, of philosophers mulling over the nature of the world. Now it was tested in the laboratory by scientists whose observing eyes and instruments took measure of our world and worlds elsewhere. In rushed marvelous advances—chemistry, gravity, the laws of motion, the basics of medicine—built the same way Euclid's *Elements* were built, with one discovery leading to another. Instead of essential natures, we were surrounded by a universe of matter, inhabited by a few inexplicable living things like us. And before we knew it, a paradigm had shifted in our understanding of the cosmos. It was as if a channel avulsed. Like the Great Nile, the sediment built, the water slowed. Before we knew it, the pressure broke, and there was a new path downstream.

With a mechanical, largely lifeless universe came a newfound existential sorrow. It meant we were potentially alone in the enormity of the now tenebrous night. But the cosmos was obtainable; we could know something about it. In this way, I am a product of my time, a captive of circumstance. I am searching the darkness because there is a universe out there awaiting discovery. It is exciting to live with such possibility, but frustrating to be aware of one's constraints. William Blake once wrote that "if the doors of perception were cleansed everything would appear to man as it is, Infinite. For man has closed himself up till he sees all things thro' narrow chinks of his cavern." That is my experience too, just a few hundred years later. We have human brains within human skulls, and we understand little of what surrounds us. The limits of our perception and knowledge are palpable, especially at the extremes, like when we're exploring space. There is so little data to tell us who we are and where we are going, why we are here, and why there is something rather than nothing. This is the affliction of being human in a time of science: We spend our lives struggling to understand, when often we will have done well, peering out through those narrow chinks, just to apprehend.

THE WORLD THAT Euclid knew—the Alexandria he once strolled—has disappeared. The city was taken by Christians, then Muslims. The lighthouse was destroyed by medieval earthquakes, the last of its remnant stones used to build a citadel. Gone are the mechanical birds that sang from the tops of trees, the steam-powered statues that raised their trumpets to the sky. Gone is the library, which was burned, its shelves emptied. None of it was permanent. It all fell away.

What parts of us—of all we know, of all we do, of all we are—will escape the same fate? Surely not the rovers, or even those Olympic rings we etched on Mars. Not our current understanding of Mars or its possibilities for life. And, on a planetary timescale, nothing on Earth, for someday the sun will die and swallow it entirely.

But it is here that I split like a river. By dint of training, experience, and circumstance, I'm with atomists, and a twenty-first-century scientist no less. I know that everything is particles and forces, that we are but a spark of light in a fundamentally inanimate universe. Just as there was a beginning of time, there was a beginning of life. And one day, there will be an end. We are unique and bounded, and we may well be in decline, for we know that species come and go. We are a finite tribe in a temporary world, marching toward our end.

And what of life itself? Must it be finite as well? What if life is a consequence of energetic systems? What if the nothing-to-something has happened time and again and, because the chinks in our cavern are so small, we don't know it? For me, this is what the search for life amounts to. It is not just the search for the other, or for companionship. Nor is it just the search for knowledge. It is the search for infinity, the search for evidence that our capacious universe might hold life elsewhere, in a different place or at a different time or in a different form. That confirmation would be a rebuke to the cratered image of Mars, the acid waters, the sterile soil. It would stand in contradistinction to the finite life to which we are confined, to the finite planet we inhabit. Finding life—even if it is the smallest microbe—would, for me, be the end of *akhet*. It would be the first dry mound emerging from the limitless dark water. An actual fact about the actual world. A truth, a beginning. It would be a shimmering hope that life might not be an ephemeral thing, even if we are.

I think of that when I pull out a certain box in my lab. Inside are articles written by Mars scientists who have long since passed away. I return to this collection again and again, with its crinkled pages and old-fashioned fonts, hand-drawn figures and hand-labeled graphs. I'll pull it down when I'm there by myself, running a script or waiting for an experiment to finish. Even though much of the science these documents postulate is not correct, or at least not entirely correct, in them I can see the great strides forward, the longing for answers.

It feels like a library of scrolls. The alluvium of my predecessors, the richness of their seeking and striving, all gathered together, and I am trying to plumb it. Next to the articles, there are copies of the riveting letters from William Pickering to his brother and the leather-bound books into which he scrawled his impressions of Mars from the patio beside a plantation house turned observatory in Jamaica. Interleaved are hundreds of pencil sketches and dozens of delicate paintings. There are stills from the silent-film footage of the transatlantic voyage during which Guglielmo Marconi tried to detect signals from Mars: In the laboratory belowdecks, headphones on, he listens intently, somber but determined, his head tilted slightly to the right as the giant aerial spins and spins. There are images of David Peck Todd, including a picture from around 1910 of him standing in an empty field next to his deflated hot-air balloon. He's wearing a long coat and driving cap and walking toward the camera. The stitched material of his balloon is caught on a tree, and his shadow falls across the collapsing bag. He doesn't know it then, but over the next fourteen years, as so poignantly captured in the photographs, he'll try over and over again to run up within shouting distance of Mars. What would he think now that the world is getting quieter? Now that we're transitioning to fiber-optic cables, now that radio broadcasts may one day cease altogether?

There's Lowell and the canals on his spiderweb maps. When he made them, he could not have known the findings of modern ophthalmology, which suggest that humans peering into the dark may catch glimpses of the faint shadows of tiny retinal veins within their own eyes. For decades, we tried so hard to make sense of those "little gossamer filaments" cobwebbing the face of Mars. Might it have been that we couldn't escape our own ghostly image?

There's a topographic map of the Asgard Range in Antarctica, a place I've flown over more than a dozen times, skimming the peaks in a Bell or AStar helicopter. Half a century has passed, and I'm still doing the same research as Wolf Vishniac, trying to detect life in one of the planet's most impossible places. Next to the map is a paper by his wife, Helen, filled with descriptions of the cells of *Cryptococcus vishniacii*—cream-colored, nonfermentive, psychrophilic, "undescribed, imperfect yeasts." Helen went on to publish numerous journal articles about her husband's cultures. She tended to the slides for decades after his death, carrying those small slips of glass with her wherever she moved, from lab to lab, until she entered assisted living a couple of years ago.

And, of course, there is a copy of *Elements,* that crowning achievement, that bygone idea. The edition I couldn't resist buying—one of the thousand editions printed since the invention of the printing press—happened to have one of the great paintings of the Romantic Era silkscreened on its cover. All the mathematics is bound by the portrait of a person standing on a precipice, caught in the wind, at once towering over the clouds and at the same time swallowed by nothingness.

This box contains the "traces of human events" that Herodotus spoke of. This is my "gift of the river." We've been wrong about many things in the search for life. It's been so hard to find an anchor, and so hard to know when our theories will no longer hold. The box reminds me of all who have come before me and what they've contributed.

It also reminds me of what is left to do. Mars, after all, is only our first step into the vast, dark night. New technologies are paving the way for life detection missions to the far reaches of our solar system, to the moons of the outer planets, far from what we once considered the "habitable zone." To worlds that hold stacks of oceans amidst shells of ice, floating like a layer cake. That spew out jets of briny water through cryovolcanoes. That have pale hills and dark rivers and hydrocarbon rain. And then there are also the planets around other stars. There could be as many as forty billion planets that could support life in the Milky Way alone, belted with moons and moonlets—potentially an entire solar system for every person on Earth. The idea of knowing these places intimately, of one day touching their surfaces, may seem ludicrous. The universe has a speed limit—it's slow, and these worlds are very far away. What could we ever know about them, besides a few details about their orbits, perhaps some spectrographic measurements of their atmospheres? They are points of light and shadow at the very edge of our sight, far beyond our grasp. Then again, that is exactly how Mars seemed only a century ago.

. . .

AS MUCH AS Mars feels like a place we understand, a place like Earth, it is still the alien other. One of my favorite things inside the box, tucked in a bent folder, is a set of pictures that *Opportunity* took in 2010. All those years ago, it seemed like such a marvel that the rover was still working. No one would have dared to believe it would have thousands more sols of science. The dust was building, the power dropping. It had been traversing the planet for six years and was already long past its ninety-day expiration date. But then a gust of wind whistled across Meridiani Planum and cleaned some of the fine particles off the solar panels. With the unexpected spike in electricity output, the team commanded the panoramic camera to take a series of pictures that could be strung together with time-lapse photography.

The flickering images captured by the rover are unforgettable. There, on an ancient plain near the equator of Mars, against an ochre sky on a dusty day, the sun is setting. A white circle of light is drifting down over the dark desert. The terrain is bare, and the sky is still in the half-light of dusk. And on the horizon, with the dust having scattered all the red light away, the sunset glows an eerie, baffling, incandescent blue.

The color makes no sense. It rattles the mind. It rips at the seams of the physical world. Scientifically, I understand it—the properties of the light, the microphysics of the system. There is no mystery to behold. And yet the mystery, like many others in our universe, is profound, nearly incomprehensible. That blue. So recognizable, yet so foreign. Shining in a halo around our shared star, calling us like a siren.

Acknowledgments

THIS IS A book about the search for life: life beyond our planet and, implicitly, a life beyond the one we live now. I have always been quick to recognize the hubris and ambition in hoping to play a role, however small, in such a pivotal breakthrough, and making it a cornerstone of my work. At times, I've worried that this kind of seeking has come with a consequence—a sense of restlessness, an inability to be content.

Not long after I finished this manuscript, the value of the life I have now, the one right here on Earth, tucked into a cozy spot in Washington, D.C., with my husband and family, was illuminated by a medical accident that pushed me to the brink of death. What was to have been an outpatient procedure landed me in an ICU on life support, with a line in my neck, a ventilator in my throat, and my mother and husband by my side. As I lay dying, a series of fourteen transfusions replaced every drop of blood in my body, allowing me to live. This book could not possibly conclude without acknowledging the fourteen anonymous people who walked into donation centers, let strangers sink needles into their arms, and then walked back into the bright day. Because of them, I have my life back, this precious life here on Earth, and I have bright days ahead.

I also feel unspeakable gratitude toward those who have gathered around me during this dark time. They include my oldest, dearest friends: Lisha and Emma and Lippy and Kayje, who traveled hundreds of miles to ease my pain, to direct my medical care, and to keep me alive, in more ways than one, and Heather, Chelsea, Ying and Ryan, Jason and Meagan, Christine, Katherine, Erin, Shayna, Maya, Stephen, Leslie, Tisha, Tessa, Maria, Antje, Sherry, Nikki, Sarah, Angus, Hannah, Lawrence and Christina, Richard and Jeannie, Meghan and Nate, Ross and Kayte, Maxine and Joel, Jacob and Patti, Ajay and Amanda, Emma and Eric, Jeff and Laura, Alan and Annie, David and Ashley, and Judy and Mike. I'm also deeply grateful for the compassionate support I received from Georgetown University and *countless* wonderful colleagues there, and, of course, my lab. Even at the low points, they were never far from my mind. This was not the first time I have been surrounded and lifted by my students and postdocs, and I know it will not be the last.

I write a lot in this book about time and scale: about how we negotiate the dissonance between geologic timescales on the one hand, and human ones on the other; about how we live out our small, shining moments as human beings on this planet, hurtling through our enormous universe. I have been blessed that my shining moment, from the day I was born, has been shared with my family in Kentucky. Although it could have been no other way, I am so thankful I was born to Kate and John, the most compelling people I've ever known. What a joy it has been to be their daughter. (And what regret I feel that my mother does not appear in this book as much as she should!) In equal measure, I could not have become the person I became without my sister, Emily—my sister darling—who brought a distilled and pure happiness into my life and in so doing, taught me the meaning of unconditional love.

I also want to thank the many grandparents and aunts and uncles and cousins who have shaped me in immeasurable ways, and I want to thank my treasured children. For the longest time, they thought there would be just one copy of this book, and it would sit on the shelf of their shared room. In a sense they were right, because these words are and will always be for them. And most profoundly, I want to thank my husband, John, whose capacious mind shines throughout this book. He read and improved every passage, just as he has improved every part of me. He has given me the big, full life I always dreamed of. I told him at our wedding that he knew my heart, and he does. It still takes my breath away that we get to explore the expanses together.

This project began simply as a collection of thoughts that would never find expression on the pages of scientific journals. It would never have become a book without Christina, who decided to publish my first nonfiction essay, remains my favorite reader, and was convinced, long before I was, that this kind of writing might interest others. The book also wouldn't have existed without sweet Matt, who, with help from Kristin, made sure I stayed in academia at the peak of my self-doubt and also introduced me to my literary agent, Jill. With her brilliant insights, Jill shaped a mess of ideas into something full of possibility, and together with Matt, led me to my editor, Amanda, who took my hand on a cold day in December and promised me it would work. She never let go. Amanda taught me everything about how to write a book. Without her help, it never would have been worthy of being printed.

Others who have supported me in this unlikely endeavor include my irreplaceable friend Dan, with his hundred years of wisdom, and Alan, with his stirring sense of the universe, as well as Kate, Tony, the Banff Centre, the Society of Fellows and the endless number of friends I made there, Marthe and Veronica and Jane for the kindness they bestowed on my children, Margaret for her interminable faith in me, the Massachusetts Cultural Council, MIT's STS Department, and the Ellen Meloy Fund and their incredible board. I am profoundly thankful for my NASA colleagues—especially Paul, Heather, Bethany, Jim, John, Melissa, Stephanie, Will, Amy and Amy, Charles, Christine, Cherie, Jen and Jen, Doug, Steelie, and others on the SAM team, Alex, Haley, Morgan, Kevin, Abby, Steve, Jack, Kate, Jamie, Chris, Eric, Andy, Lee, Mary Beth, Jen, Dale, Tori, Brit, Joe, Paul, Lindsay, and Mary—to the early Mars mission scientists and engineers—Norm, John, Ben, Larry, and Gentry—and to my unsurpassed mentors—Maria, Ray and Eloise, Rick, Jim, Steve, John, Shere, Lindy, Pete, Gary, Scott, Dave, Roger, Kathy, Barb, Charles, Eske, Mark, and Roz. I am deeply grateful to Bill, who enhanced the manuscript tremendously with his historical depth, and Parker, who meticulously fact-checked each line, as well as Zach, James, Julie, Owen, Anita, Katie, Anne Cat, Maya, and Matt, and the staff of the Caltech, Harvard, MIT, Oxford, and JPL archives, the NASA History Office, and the Library of Congress, who all helped in various ways with research. In addition, I will always be indebted to those who read the leaden text of my early drafts—dearest Liz and Greg, Dedi, Maura and Heidi, and of course Deirdre, who is not only one of my closest friends, but also the reason I found and fell in love with John. I apologize to anyone I've inadvertently omitted, and also to those who have helped me in ways I'm not even aware of.

Eudora Welty once wrote about how things are often too indefinite to be recognized for themselves, to connect into a larger shape, until you are almost close enough to grasp them. But then "suddenly a light is thrown back, as when your train makes a curve, showing that there has been a mountain of meaning rising behind you on the way you've come, is rising there still." In writing this book, I've come to understand better the meaning I find in searching for life. I've also come to appreciate all the people who came down this path before me and the astonishing lives they led, as well as the remarkable colleagues with whom I have the privilege of working today. In my final acknowledgments, I wish to extend my gratitude to all of those people, throughout the generations and across the disciplines, who have created and continue to deepen this field. If we find life on Mars, we will have done it together. In the meantime, we have this great human project, and we have one another.

Notes

Prologue

ix WORLD'S OLDEST ROCKS J. R. De Laeter, I. R. Fletcher, K. J. R. Rosman, et al., "Early Archaean Gneisses from the Yilgarn Block, Western Australia," *Nature,* 292 (1981), pp. 322–324; D. R. Mole, M. L. Fiorentini, N. Thébaud, et al., "Archean Komatiite Volcanism Controlled by the Evolution of Early Continents," *Proceedings of the National Academy of Sciences,* 111 (June 2014).

ix CORROSIVE AS BATTERY ACID The pH levels in these unique acid salt lakes, which have been studied for years as a Mars analog because of the pioneering work of West Virginia University Professor Kathy Benison, have been recorded to fall as low as 1.6. For more about the lakes and their extreme geochemical conditions, see: K. C. Benison and D. A. LaClair, "Modern and Ancient Extremely Acid Saline Deposits: Terrestrial Analogs for Martian Environments?" *Astrobiology,* 3, no. 3 (2003), pp. 609–618; B. B. Bowen and K. C. Benison, "Geochemical Characteristics of Naturally Acid and Alkaline Saline Lakes in Southern Western Australia," *Applied Geochemistry,* 24 (2009), pp. 268–284; S. S. Johnson, M. G. Chevrette, B. L. Ehlmann, and K. C. Benison, "Insights from the Metagenome of an Acid Salt Lake: The Role of Biology in an Extreme Depositional Environment," *PLOS One,* 10 (April 2015).

x HOPING TO SIGNAL MARS Hans Zappe, *Fundamentals of Micro-Optics,* 1st ed. (Cambridge University Press, 2010), p. 298; Louise Leonard, *Percival Lowell, An Afterglow* (Boston: Richard G. Badger, 1921).

x NO LAKES There are no longer any traditional lakes (i.e., lakes on the surface on Mars), but in 2018, the MARSIS Instrument on the European Space Agency's

Mars Express orbiter found intriguing evidence of a twenty-kilometer-wide subglacial lake deep under the south polar layered deposits. See: R. Orosei, S. E. Lauro, E. Pettinelli, et al., "Radar Evidence of Subglacial Liquid Water on Mars," *Science*, 3 (Aug. 2018), pp. 490–493.

x NO PLATE TECTONICS Many modelers agree that even an early epoch of plate tectonics on Mars is difficult to reconcile with the evidence for early crust formation and magnetic-field generation, though it has been suggested that Valles Marineris could be a plate boundary. See: D. Breuer and T. Spohn, "Early Plate Tectonics Versus Single-Plate Tectonics on Mars: Evidence from Magnetic Field History and Crust Evolution," *Journal of Geophysical Research: Planets*, 108, no. E7 (2003); An Yin, "Structural Analysis of the Valles Marineris Fault Zone: Possible Evidence for Large-Scale Strike-Slip Faulting on Mars," *Lithosphere*, 4, no. 4 (2012), pp. 286–330.

x NO MAGNETIC FIELD Most estimates suggest that a global magnetic field of appreciable magnitude dissipated around four billion years ago. See: David J. Stevenson, "Mars' Core and Magnetism," *Nature*, 412, no. 6843 (2001), p. 214; Sean C. Solomon, Oded Aharonson, Jonathan M. Aurnou, W. Bruce Banerdt, Michael H. Carr, Andrew J. Dombard, Herbert V. Frey, et al., "New Perspectives on Ancient Mars," *Science*, 307, no. 5713 (2005), pp. 1214–1220; and J. E. P. Connerney, J. Espley, P. Lawton, S. Murphy, J. Odom, R. Oliversen, and D. Sheppard, "The MAVEN Magnetic Field Investigation," *Space Science Reviews*, 195, no. 1–4 (2015), pp. 257–291.

xi MUCH MORE LIKE EARTH Mars's diameter is a bit more than half as large as Earth's; in comparison, you would need more than eleven Earths side by side to match the diameter of Jupiter.

xi LIFTED GREENHOUSE GASES See: R. M. Haberle, "Early Mars Climate Models," *Journal of Geophysical Research*, 103 (Nov. 1998), pp. 28,467–28,479; I. Halevy, M. T. Zuber, and D. P. Schrag. "A Sulfur Dioxide Climate Feedback on Early Mars," *Science* 318, no. 5858 (2007), pp. 1903–1907; S. S. Johnson, M. A. Mischna, T. L. Grove, and M. T. Zuber, "Sulfur-Induced Greenhouse Warming on Early Mars," *Journal of Geophysical Research: Planets*, 113, no. E8 (2008); R. M. Ramirez, et al., "Warming Early Mars with CO2 and H2," *Nature Geoscience*, 7 (2014), pp. 59–63; and R. D. Wordsworth, "The Climate of Early Mars," *Annual Review of Earth and Planetary Sciences*, 44 (2016), pp. 381–408.

xi WARM AND WET, AT LEAST PERIODICALLY See: R. A. Craddock and A. D. Howard, "The Case for Rainfall on a Warm, Wet Early Mars," *Journal of Geophysical Research Planets*, 107 (Nov. 2002), pp. 21–36; S. W. Squyres and J. F. Kasting, "Early Mars: How Warm and How Wet?" *Science*, 265 (Aug. 1994); R. D. Wordsworth, et al., "Comparison of 'Warm and Wet' and 'Cold and Icy' Scenarios for Early Mars in a 3-D Climate Model," *Journal of Geophysical Research Planets*, 120 (June 2015), pp. 1,201–1,219; M. C. Palucis, et al., "Sequence and Relative Timing of Large Lakes in Gale Crater (Mars) after the Formation of Mount Sharp," *Journal of Geophysical Research: Planets*, 121, no. 3 (2016), pp. 472–496.

xi "WARM LITTLE PONDS" Charles Darwin, letter to J. D. Hooker, February 1, 1871, Darwin Correspondence Project. In the letter to Charles Hooker, Darwin wrote: "But if (and oh what a big if) we could conceive" of the origin of life "in some warm little pond . . ." His instincts may have been right. Two leading possibilities today for the locus of the origin of life on Earth are hydrothermal vents deep in the ocean and freshwater pools within geothermal fields, similar to those in Yellowstone National Park. The latter has come to prominence in recent years for several reasons. The chemical composition of cells more closely resembles the chemical composition of pooling water within geothermal fields than that of deep-sea waters; organic molecules—the building blocks of life—could have accumulated in ponds more easily than in the deep ocean, and lower salt concentrations could have provided a more conducive environment for the first fatty-acid membranes to form. In addition, recent work now suggests that repeated cycles of wetting and drying may have been necessary to pattern the precursors of repeating informational molecules in membranous vesicles. If land is indeed required for life, at least life as we know it, Mars may hold more possibility than the icy moons of Jupiter and Saturn. See: Armen Y. Mulkidjanian, et al., "Origin of First Cells at Terrestrial, Anoxic Geothermal Fields," *Proceedings of the National Academy of Sciences*, 109 (2012), pp. E821–E830; D. Deamer and B. Deamer, "Can Life Begin on Enceladus? A Perspective from Hydrothermal Chemistry," *Astrobiology* (Sept. 2017), pp. 834–839; D. Deamer, *First Life: Discovering the Connections between Stars, Cells, and How Life Began* (Berkeley: University of California Press, 2011).

xi NORTHERN OCEAN Articles about the possibility of an ancient ocean on Mars include: M. H. Carr and J. W. Head III, "Oceans on Mars: An Assessment of the Observational Evidence and Possible Fate," *Journal of Geophysical Research Planets*, 108 (2003); R. I. Citron, M. Manga, and D. J. Hemingway, "Timing of Oceans on Mars from Shoreline Deformation," *Nature*, 555 (2018), pp. 643–646; G. Di Achille and B. M. Hynek, "Ancient Ocean on Mars Supported by Global Distribution of Deltas and Valleys," *Nature Geoscience*, 3 (2010), pp. 459–463; M. C. Malin and K. S. Edgett, "Oceans or Seas in the Martian Northern Lowlands: High Resolution Imaging Tests of Proposed Coastlines," *Geophysical Research Letters*, 26 (1999), pp. 3,049–3,052.

xi ABYSSAL PLAINS J. W. Head III, et al., "Oceans in the Past History of Mars: Tests for Their Presence Using Mars Orbiter Laser Altimeter (MOLA) Data," *Geophysical Research Letters* (Dec. 1998), p. 4,403; J. W. Head III, et al., "Possible Ancient Oceans on Mars: Evidence from Mars Orbiter Laser Altimeter Data," *Science*, 286 (1999), pp. 2,134–2,137.

xi THREE AND A HALF J. P. Bibring, et al., "Global Mineralogical and Aqueous Mars History Derived from OMEGA/Mars Express Data," *Science*, 312 (April 2006), pp. 400–404.

xi ALMOST ALL OF THE ATMOSPHERE With essentially no greenhouse effect, the surface temperatures of Mars, following the Stefan-Boltzmann law, slowly dropped to an average of minus 60 degrees Celsius, the surface temperature today.

xi DUST THE CONSISTENCY OF K. S. Edgett and H. E. Newsom, "Dust Deposited from Eolian Suspension on Natural and Space Flight Hardware Surfaces in Gale Crater as Observed Using Curiosity's Mars Hand Lens Imager (MAHLI)," presented at Dust in the Atmosphere of Mars and Its Impact on Human Exploration, Houston, Texas (June 2017).

xii JUST ONE OF THE MANY FEATURES Where we should look and what we should look for is a matter of continual debate. Astrobiologists bump heads about the definition of life—and whether a definition even makes sense. See: Benton Clark, "A Generalized and Universalized Definition of Life Applicable to Extraterrestrial Environments," in *Handbook of Astrobiology*, ed. Vera M. Kolb (Boca Raton, Florida: CRC Press, 2018), and for an alternative line of thinking, see: C. E. Cleland and C. F. Chyba, "Defining Life," *Origins of Life and Evolution of Biospheres* 32 (2002), pp. 387–393; C. E. Cleland and C. F. Chyba, "Does 'Life' Have a Definition?" in *Planets and Life: The Emerging Science of Astrobiology*, ed. W. T. Sullivan III, and J. A. Baross (Cambridge: Cambridge Univ. Press, 2007), pp. 119–131; C. E. Cleland, *The Quest for a Universal Theory of Life: Searching for Life As We Don't Know It* (Cambridge: Cambridge Univ. Press, 2019). Philosopher Carol Cleland argues that, lacking a theory of life, trying to come up with a definition of life is premature. Water, for instance, was once described with terms like "wet" and "thirst-quenching" but couldn't be meaningfully defined until the existence of molecular theory—until the discovery of hydrogen and oxygen atoms.

xiii AS A RESULT, MARS While humans have analyzed meteorites from Mars that have landed on Earth, no humans have ever landed on the planet's surface.

Chapter 1: Into the Silent Sea

3 KÁRMÁN LINE Because Earth's atmosphere becomes thinner with increasing altitude, it is difficult to say precisely where space begins. The Hungarian-American scientist Theodore von Kármán initially described the border between Earth's atmosphere and outer space to be approximately eighty kilometers above sea level—"where aerodynamics stops and astronautics begins." Today the Fédération Aéronautique Internationale defines the Kármán line as one hundred kilometers above sea level. Where space begins has far-reaching implications for space policy and will likely become increasingly important as commercial space-flight ventures increase suborbital activity.

4 "TODAY THE FINGERTIP" Walter Sullivan, "Mankind, Through Mariner, Reaching for Mars Today," *The Courier-Journal* (Louisville, Ky., July 14, 1965).

5 VON KÁRMÁN AUDITORIUM Ray Duncan, "Army of Newsmen to Jam Pasadena for Mars Probe," *The Independent* (Pasadena, Calif., July 12, 1965).

5 "EARLY BIRD" Ibid.

5 THIRTY-SEVEN PHONES Ibid.

5 TEMPERATURE-CONTROL TESTING Ibid.

5 138,000 PARTS J. N. James, "The Voyage of Mariner IV," *Scientific American*, 214, no. 3 (March 1966), pp. 42–53.

5 "VOICELESS 'RUSSIAN SPY'" Dave Swaim, "Mars Spaceship Has Company," *The Independent* (Pasadena, Calif., July 13, 1965).

5 "'DEAD' SOVIET MARS" "'Dead' Soviet Mars Missile Still on Way," *Pasadena Star-News* (Pasadena, Calif., July 14, 1965).

5 310 WATTS Edward Clinton Ezell and Linda Neuman Ezell, "On Mars: Exploration of the Red Planet, 1958–1978" (Washington, D.C.: The NASA History Series, 1984), p. 434.

5 TENTH OF A BILLIONTH OF A BILLIONTH "To Mars: The Odyssey of Mariner IV," JPL Technical Memorandum, No. 33-229, p. 24.

6 GREAT DISHES *Mariner 4* was tracked twelve hours a day by the Goldstone station in California; the remaining time was covered by the Australian station at Tidbinbilla (outside Canberra) and a station at Johannesburg, which has since been replaced by a Spanish complex sixty kilometers west of Madrid. "To Mars: The Odyssey of Mariner IV," p. 25; Douglas J. Mudgway and Roger Launius, "Uplink-Downlink: A History of the Deep Space Network, 1957–1997" (Washington, D.C.: The NASA History Series, 2001).

6 THERE WERE WORRIES Blaine Baggett, dir., *The Changing Face of Mars: Beginnings of the Space Age* (Pasadena, Calif.: Jet Propulsion Laboratory, 2013), DVD video.

6 A FORMIDABLE ADVERSARY The Soviets were extremely secretive about their space program. While *Mars 1,* the mission that made it 106 million kilometers before the radio transmitter went dead, was announced at the time, others were long matters of speculation in the West.

6 THE FIRST SENTIENT BEINGS The first being sent into orbit was Laika the dog, a small stray mongrel from the streets of Moscow, who launched on *Sputnik 2* on November 3, 1957. Laika died, likely of overheating, just a few hours after launch, and the capsule burned up on reentering the Earth's atmosphere on April 14, 1958. The first creatures to return safely were launched on *Sputnik 5,* or *Korabl-Sputnik 2,* on August 19, 1960. The dogs were named Belka and Strelka ("Whitey" and "Little Arrow"). They were taxidermized after dying of old age. Their bodies now reside at the Museum of Cosmonautics in Moscow.

6 NOT SO LUCKY Andrew J. LePage, "The Beginnings of Planetary Exploration," *The Space Review* (October 11, 2010).

6 ROCKET TO MARS FAILED Caleb A. Scharf, "The Long Hard Road to Mars," *Scientific American,* Nov. 25, 2011; "Marsnik 2," NASA, Space Science Data Coordinated Archive, Sept. 5, 2019.

6 PULL OFF HIS SHOE This has been widely rumored but remains an unconfirmed story. See: William Taubman, "Did He Bang It? Nikita Khrushchev and the Shoe," *The New York Times* (July 26, 2003); C. Eugene Emery Jr., "The Curious Case of the Khrushchev Shoe," *PolitiFact* (Jan. 18, 2015); Arthur I. Cyr, "PolitiFact Bizarrely, Unjustly Attacks Me on Krushchev Shoe Banging," *Providence Journal* (Feb. 22, 2015).

6 A MUTE WITNESS At the time, long-range radio communications were the Soviets' Achilles' heel, one area where the Americans had the advantage. For a terrific resource on the early Soviet space missions, see: James Harford, *Korolev: How One Man Masterminded the Soviet Drive to Beat America to the Moon* (New York: John Wiley & Sons, Inc., 1997).

7 THEIR SUCCESSES The first spacecraft that crashed into the moon was *Luna 2,* in September 1958, and the first to take pictures of the far side—an incredible feat at the time—was *Luna 3,* in October of the same year. The first spacewalk, completed by Soviet cosmonaut Alexei Leonov on *Voskhod 2,* occurred in March 1965, and it preceded Ed White's spacewalk on *Gemini 4* by just under three months.

7 FIRST ARTIFICIAL SATELLITE *Sputnik* was launched on the same vehicle as the first ICBM, and just six weeks later.

7 MARINER 2 TO VENUS *Mariner 1* was also designed to fly by Venus. It launched on July 22, 1962, on an Atlas-Agena rocket. A DESTRUCT command was sent shortly thereafter, after launch data indicated the mission was doomed.

7 "MISSION OF SEVEN MIRACLES" Franklin O'Donnell, "The Venus Mission: How Mariner 2 Led the World to the Planets," NASA, JPL/Caltech (2012).

7 "LIMPING ON ONE SOLAR PANEL" Ibid.

7 GETTING TO VENUS Launch windows for Venus occur more frequently than launch windows for Mars, once every 584 days compared to 780 days. Because Venus is closer, the flight time is also shorter, and communications need to travel a shorter distance.

7 NEVER-BEFORE-TESTED STAR TRACKER Norman Haynes, personal interview by Sarah Johnson (Pasadena, Calif., Feb. 1, 2018).

7 "THE FIRST THAT EVER BURST" Samuel Taylor Coleridge, *The Rime of the Ancient Mariner* (Poetry Foundation, 1834 text).

7 TWO *MARINER* MISSIONS TO MARS In 1960, NASA decided that lunar probes would be named after land-exploration activities, whereas planetary-mission probes would be named after nautical terms, to connote "the impression of travel to great distances and remote lands." For more on naming conventions, see: Helen T. Wells, Susan H. Whiteley, and Carrie E. Karegeannes, "Origins of NASA Names" (Washington, D.C.: The NASA History Series, 1976).

7 "FLYING WINDMILLS" Baggett, *The Changing Face of Mars*.

7 THE NOSE FAIRING "Mariner 4 Probe Due in Two Weeks," *Pasadena Star-News* (Nov. 6, 1964).

8 DERELICT ORBIT AROUND John C. Waugh, "Mars Probe Falls Silent," *The Christian Science Monitor* (Boston: Nov. 1964).

8 MARS AND EARTH ALIGN These periods, when Mars and the sun align on opposites sides of the Earth—when Mars and Earth are closest—are called oppositions.

8 POP LIKE POPCORN Haynes, personal interview by Johnson.

8 POKING TINY HOLES Ibid.

8 PAD 37 Jack N. James, *In High Regard* (Jack James Trust, 2006), p. 450.

8 DESIGNED TO DO "Mars Flight—on Up & Up," *Pasadena Star-News* (Nov. 28, 1964).

8 CLUSTER OF THREE STARS Marvin Miles, "Mariner 4 Locks on to Key Star After Four Misses," *Los Angeles Times* (Dec. 1, 1964); Marvin Miles, "Mariner to Fly Within 5,400 Miles of Mars," *Los Angeles Times* (Dec. 11, 1964).

8 KEPT GETTING DISLODGED Initially, the gyro control unit could only be turned on by ground control if Canopus was lost, but a command sent on December 17, 1964, automated the reacquisition. "Mariner Mars 1964 Project Report: Spacecraft Performance and Analysis," JPL Technical Report, No. 32-882 (Pasadena, Calif.: NASA, 1967), p. 17; "To Mars: The Odyssey of Mariner 4," pp. 21–22; W. C. Goss, "The Mariner Spacecraft Star Sensors," *Applied Optics,* 9, issue 5 (1970), pp. 1,056–1,067.

8 HALFWAY AROUND THE SUN Earth and Mars are revolving around the sun, so as a spacecraft travels to Mars, it also revolves around the sun.

9 "PICTURES, THAT'S NOT SCIENCE" This quote is attributed to Bud Schurmeier, though he did not share this view. "Mariner 4 Taught Us to See," JPL Blog (August 20, 2013).

9 BUT LEIGHTON HAD A DEEP PASSION For a detailed recounting of Leighton's early life and involvement in the early *Mariner* missions, see: Heidi Aspaturian, "Interview with Robert Leighton," *California Institute of Technology Oral History Project (1986–1987)*, California Institute of Technology Archives and Special Collections (Pasadena, Calif., 1995).

9 LEIGHTON HAD SPENT THE DECADE Ibid., p. 64.

9 ONE OF LEIGHTON'S STUDENTS The student was Gerry Neugebauer, who went on to play a key role in the development of infrared planetary astronomy.

9 TOLD HIS COLLEAGUES Bruce Murray, interviewed by Rachel Prud'homme, audiocassette recording transcript, California Institute of Technology Archives and Special Collections (Pasadena, Calif., 1993), p. 76.

10 "BOB, AS A DUTY TO" Graham Berry, "Interview with Robert P. Sharp," California Institute of Technology Archives and Special Collections (Pasadena, Calif., 2001), p. 43.

10 TWENTY-ONE PHOTOGRAPHS "Press Kit, Mariner Mars Encounter," NASA (July 9, 1965).

10 THE HOPE WAS TO COLLECT It was lucky the actual images (which captured Amazonis Planitia, Elysium, Western Memnonia Fossae, Eastern Gorgonum Chaos, Orcus Patera, and Aonia Terra) came as close to the hoped-for path as they did, as even a tiny midcourse error would have moved the spacecraft by hundreds of kilometers. Intended path from "Press Kit, Mariner Mars Encounter," NASA.

10 WORLD'S FIRST DIGITAL CAMERA The first digital camera with a CCD image sensor wasn't developed at Eastman Kodak until the mid-1970s, but *Mariner 4*'s camera was the first to transmit images in binary code—i.e., as a long series of zeros and ones. See: Fred C. Billingsley, "Processing Ranger and Mariner Photography," *Optical Engineering,* 4, no. 4, 404147 (May 1, 1966); "First Digital Image From Space (Mariner 4-Mars)," NASA, JPL/Caltech (Sept. 28, 2018).

10 "SCIENTIFIC PRIEST" Murray, interviewed by Prud'homme, p. 79.

10 THE "FIXED" STARS Enn Kasak and Raul Veede, "Understanding Planets in Ancient Mesopotamia," *Folklore,* 16 (2001).

10–11 BLAZING RED LAMP Mars can vary dramatically in brightness. Although Jupiter is usually brighter (even at perihelic oppositions), during the August 2003 opposition, when Mars was closer than it had been since the Neanderthals still controlled Europe, it outshone every other object in the night sky except the moon and Venus.

11 ELONGATED LOOP In Babylon, the planets were regarded as the "interpreters" of the gods, leading priest-astronomers to carefully archive records of the planets' motions. Later, as the arithmetic methods of the Babylonians were blended

with the spatial and geometric imagination of the Greeks, the planets were seen as moving in the circles-on-circles of epicycles around the Earth, which was the hearth at the center of the whole system. Mars rode on a large epicycle, which was needed to explain the large loops backward that it underwent every two years, and since the size of its loops varied, the center of its epicycle had to be placed off-center to the Earth. By the sixteenth century, Copernicus decided to put the sun at the center of his system, and Mars's largest epicycle was no longer needed—since it was now evident that it was merely an illusion of perspective, created as the Earth, following its own orbit around the sun, caught up with and passed the slower-moving Mars. In 1609, Johannes Kepler worked out the elliptic shape of the planetary orbits, using the orbit of Mars, which is highly elliptical, as the basis of his protracted and difficult calculations. This was one of the great discoveries of modern astronomy—and we owe it all to Mars.

11 PLATO CONCLUDED THAT Plato, "Book X," *The Republic,* trans. Benjamin Jowett (Cambridge: Internet Classics Archive, MIT, 2008). See also the claim that, based on the planets' observed "retrogradation" and "wobbling," "it almost seemed (for example, to Plato when the *Timaeus* was being written) that nothing less than an exercise of free will could account for their reversals of direction": Robert Sherrick Brumbaugh, *Plato for the Modern Age* (Lanham, Md.: University Press of America, 1991).

11 PERSPICILLUM, OR TELESCOPE Although the principle of the telescope had been discovered by lens makers in Holland a year or two earlier, Galileo independently worked out everything else, and he built several telescopes on his own, of which only two survive. The better of these has a five-centimeter lens and is about a meter long, with a tube consisting of two grooved wooden shells held together by copper bands and covered in paper. For more information, see Giorgio Strano, ed., *Galileo's Telescope: The Instrument that Changed the World* (Florence, Italy: Istituto e Museo di Storia della Scienza, 2008).

11 UTTERLY STILL David Wootton, *Galileo: Watcher of the Skies* (New Haven: Yale University Press, 2010), p. 96.

11 THROUGH ITS TINY APERTURE Galileo didn't actually get around to Mars until later in 1610; with a telescope, he also discovered craters and mountains on the moon, the four large satellites of Jupiter, and the stars of the Milky Way.

11 A SPHERICAL BODY William K. Hartmann and Odell Raper, *The New Mars: The Discoveries of Mariner 9* (Washington, D.C.: NASA Office of Space Science, 1974), p. 1; Galileo even suspected—though this was at the very limit of what he could make out—that it showed a phase, making it look like a tiny gibbous moon.

11 "IF WE COULD BELIEVE" Galileo Galilei, "Third Letter on Sunspots, from Galileo Galilei to Mark Welser, In which Venus, the Moon, and the Medicean Planets Are also Dealt With, and New Appearances of Saturn Are Revealed," *Discoveries and Opinions of Galileo,* Stillman Drake, ed. (New York: Anchor Books, 1957), p. 137.

11 POPPY SEED Galileo's telescope, the best in the world at the time, magnified twenty times, but Mars was far away from Earth when Galileo first observed it.

11 CONCAVE LENSES Such as the ones Galileo used as eyepieces in his first telescopes.

11 IN 1659 This night in 1659, November 28, was three hundred five years to the day before *Mariner 4* launched to Mars.

12 BLOB DISAPPEAR AND REAPPEAR William Sheehan, *The Planet Mars: A History of Observation and Discovery* (Tucson: University of Arizona Press, 1999), p. 21.

12 HOURGLASS SEA This feature is now known as Syrtis Major; the dark color comes from dust-free basaltic rock.

12 60 PERCENT THE SIZE The true proportion is 53 percent.

12 ROUGHLY TWENTY-FOUR HOURS The Italian astronomer Giovanni Cassini, another leading observer of Mars at the time, soon modified this to twenty-four hours and forty minutes.

12 "NOTHING BUT VAST DESERTS" Christiaan Huygens, "Cosmotheoros," quoted in William Miller, *The Heavenly Bodies: Their Nature and Habitability* (London: Hodder and Stoughton, 1883), p. 101.

12 "THERE'S NO REASON" Christiaan Huygens, *The Celestial Worlds Discover'd: Or, Conjectures Concerning the Inhabitants, Plants, and Productions of the Worlds in the Planets* (London: Timothy Childe, 1698; Digitized by Utrecht University).

12 A HAZE OF COLOR This is known as chromatic aberration.

12 THIRTY-FIVE TIMES Isaac Newton, *Opticks: Or, A Treatise of the Reflections, Refractions, Inflections, and Colours of Light* (London: William and John Innys at the West End of St. Paul's, 1721), p. 91.

12 WITHIN A CENTURY In the meantime, with some desperation, astronomers were trying to mitigate the problem of chromatic aberration with increasingly long refracting telescopes. Huygens tried dispensing with the telescopic tube altogether, mounting a lens on a high mast, and controlling it by means of a guy wire tethered to the eyepiece holder near the ground. An assistant would hold out a lantern to illuminate the lens while Huygens searched for its reflection and adjusted the focus. Huygens experimented with "aerial" telescopes that were 37, 52, and 64 meters in length. They took ages to align and were awash in stray light, and despite the extraordinary lengths to which he went, his observations of Mars did not improve significantly. Cassini, meanwhile, was setting up 30- and 41-meter focal-length telescopes, mounted atop an old wooden tower. The tower was equipped with a stairway and a balcony around the top to prevent his assistants from falling off in the darkness. See: Sheehan, *The Planet Mars*, pp. 24–26.

12 ENABLING UNPRECEDENTED ENLARGEMENTS It was not until 1722 that John Hadley produced a reflector that was as good as Huygens's aerial telescopes. By then the refractor was on the verge of making a comeback, with the discovery of achromatic lenses. These compound lenses, where a concave lens of flint glass was used in combination with a convex lens of crown glass, allowed the chromatic aberration produced by one lens to be largely compensated for by the other (at least within a certain range of wavelengths, including the yellow light where the eye is most sensitive). Good achromatic lenses were coming into use by the middle of the eighteenth century, but they were expensive, and it was their expense that led Herschel to build reflectors.

12 YOUNGER SISTER, CAROLINE Caroline was also a highly accomplished astronomer. She was an honorary member of the Royal Astronomical Society and

was presented with a Gold Medal for Science at the age of 96 by the king of Prussia.

13 WHITE POLAR CAPS Giovanni Cassini and Christian Huygens also observed polar caps.

13 "CLOUDS AND VAPORS" William Herschel, quoted in *The New Mars: The Discoveries of Mariner 9,* p. 2.

13 "SIMILAR TO OUR OWN" William Herschel, quoted in Chris Impey and Holly Henry, *Dreams of Other Worlds: The Amazing Story of Unmanned Space Exploration,* rev. ed. (Princeton, N.J. Princeton University Press, 2016), p. 15.

15 "LOOKING AT A PLANET" Murray, interviewed by Prud'homme, p. 162.

15 VIA TELETYPE Baggett, *The Changing Face of Mars.*

15 WERE LIKE PEARLS Aspaturian, "Interview with Robert Leighton," p. 103.

15 THE DATA RATE Baggett, *The Changing Face of Mars.*

15 EIGHT HOURS "To Mars: The Odyssey of Mariner 4," p. 30.

15 HAD DECIDED TO RELAY Baggett, *The Changing Face of Mars.*

16 CONTEST OF SORTS Baggett, *The Changing Face of Mars.*

16 GRUMM POPPED OVER Dan Goods, "First TV Image of Mars," DirectedPlay .com.

17 *VIDIMUS ET ADMIRATI SUMUS* Brandon A. Evans, "What Was in the News on July 23, 1965?" *The Criterion,* online edition (July 24, 2015); Latin translation courtesy of Charlayne Allan.

17 STREAKS ACROSS THE FRAME AP, UPI, and *L.A. Times–Washington Post* dispatches; "Mariner 4 Shot Shows Mars Hills," *The Courier-Journal* (Louisville, Ky., July 16, 1965).

17 "THE RESOLUTION WAS AWFUL" John Casani, personal interview by Sarah Johnson (Pasadena, Calif.; Aug. 6, 2015).

17 "JACK, YOU AND I" James, *In High Regard,* p. 456.

17 "IT'S THE MOON" Haynes, personal interview by Johnson.

18 HIS INAUGURAL SPEECH Lyndon B. Johnson, "The President's Inaugural Address, January 20, 1965," Gerhard Peters and John T. Woolley, eds., The American Presidency Project (University of California, Santa Barbara).

18 IN THE EAST ROOM Lyndon B. Johnson, *Public Papers of the Presidents of the United States: Lyndon B. Johnson, 1965* (Best Books, 1965), pp. 805–806.

18 "A PROFOUND FACT" Robert B. Leighton, "Mariner 4 Press Conference," eFootage.com (July 29, 1965).

18 "IT MAY BE" Johnson, *Public Papers of the Presidents,* pp. 805–806; Baggett, *The Changing Face of Mars.*

18 INSTRUMENTS ALSO REVEALED "Mariner 4," NASA Space Science Data Coordinated Archive, NSSDCA/COSPAR ID: 1964-077A.

19 "CRATERS? WHY DIDN'T WE THINK" Oliver Morton, *Mapping Mars: Science, Imagination, and the Birth of a World* (New York: Picador, 2002), p. 73.

19 "A DEAD PLANET" "The Dead Planet," *The New York Times* (July 30, 1965).

Chapter 2: The Light that Shifts

20 IN THE IMAGE "Mariner 4's First Picture Clearly Showing Craters on Mars," NASA, JPL/Caltech (1965).

20 OF THE U.S. AIR FORCE Betsy Mason, "What Mars Maps Got Right (and Wrong) Through Time," *National Geographic* (Oct. 16, 2016).

20 PEACH AND GRAY Aeronautical Chart and Information Center, "Mars: MEC-1 Prototype," Library of Congress (1965).

21 SIX IN ALL Ibid.

21 PULL-APART ORIGIN Gregory A. Davis, "2009 Penrose Medal Presented to B. Clark Burchfiel, Citation by Gregory A. Davis," The Geological Society of America (2009).

22 SHILLING A MILE Beryl Markham, *West with the Night: A Memoir* (New York: North Point Press, 2013), p. 198.

22 HEMINGWAY CALLED THE BOOK Diane Ackerman, "A High Life and a Wild One," *The New York Times* (Aug. 23, 1987).

22 LADISLAUS DE ALMÁSY Ondaatje's protagonist was inspired by the historical figure László Almásy. Michael Ondaatje, *The English Patient* (New York: Vintage Books, 1993), p. 16.

23 SHEEN POTENTIALLY LINKED The mechanism for the formation of desert varnish remains a topic of debate; see: Naama Lang-Yona, et al., "Insights into Microbial Involvement in Desert Varnish Formation Retrieved from Metagenomic Analysis," *Environmental Microbiology Reports,* 10, no. 3 (June 2018), pp. 264–271; Phil Berardelli, "Solving the Mystery of Desert Varnish," *Science* (July 7, 2006).

23 "FACE TO FACE WITH A HATCHERY" Antoine de Saint-Exupéry, *Wind, Sand and Stars* (Boston: Harcourt, 2002), p. 111.

23 "SEIZED WITH VERTIGO" Ibid.

23 THE MARS PLANNING MAP I first discovered the map through Emily Lakdawalla's wonderful blog post, "Mapping Mars, Now and in History," The Planetary Society (Feb. 26, 2009). This map and other fascinating photos, videos, artwork, diagrams, and amateur-processed space images can be found in the Planetary Society's online Bruce Murray Space Image Library.

24 DISCOVERED TWO MOONS The astronomer was Asaph Hall. George William Hill, *Biographical Memoir of Asaph Hall, 1829–1907* (Washington, D.C.: The National Academy of Sciences, 1908), pp. 262–263.

24 SCHIAPARELLI'S TELESCOPE The Brera Observatory in Milan was an old observatory whose equipment was largely antiquated when Schiaparelli arrived. However, a close associate of one of the king of Italy's ministers, who had studied engineering with Schiaparelli at the University of Turin, persuaded the Ital-

ian parliament to fund a new telescope. It was delivered in 1875. Schiaparelli's first priority was to use it to observe double stars, for which it was very well suited.

24 HIS NOTEBOOK Agnese Mandrino, et al., "Ed ecco Marte!" *Di Pane e Di Stelle* (April 5, 2010); G.V. Schiaparelli, "First observations of Mars: Thursday, August 23, 1877," Notebook Entry, *Historical Archive of the Astronomical Observatory of Brera,* Box 403: 1 and Box 407: 1.

24 TWO CARTOON HANDS R. A. Proctor, "Proctor's Mars Maps (1865–1892)," *Planetary Maps* (Jan. 29, 2016).

24–25 A TINY MICROMETER Schiaparelli developed a great deal of experience using a micrometer to measure double stars as a student of Otto Wilhelm Struve and Johann Encke (in the decades between the time Schiaparelli began observing double stars and the time he gave up, owing to failing eyesight, he made thousands of double-star measurements).

25 PRETERNATURALLY SHARP Schiaparelli was also color-blind, meaning that he was probably more sensitive to slight gradations of intensity of the markings such as those between boundaries of different tone.

25 "SALTIER THE WATER" David A. Weintraub, *Life on Mars: What to Know Before We Go* (Princeton, N.J.; Princeton University Press, 2018), p. 91.

25 OF THESE "CANALI" Schiaparelli borrowed the term "canali" from another Italian astronomer, who had dubbed a dark space the "Atlantic Canale" some fifteen years earlier, for it seemed to separate two bright continents.

25 MIDDLE OF A LANDMASS Weintraub, *Life on Mars,* p. 93.

25 CAMILLE FLAMMARION Flammarion was also a mystic. In the late Victorian Era, as intellectual life was wresting itself from the authority of the church, his blend of science and spiritualism was a peculiar, ephemeral compromise. Flammarion made no distinction between life after death and life on other worlds within the observable universe. He thought of humans as "citizens of the sky." He believed that human souls passed from planet to planet and that telepathy was "as much a fact as London, Sirius, and oxygen." Camille Flammarion, *Camille Flammarion's The Planet Mars,* trans. Patrick Moore (New York: Springer, 2014); Robert Crossley, "Mars and the Paranormal," in *Imagining Mars: A Literary History,* (Middletown, Conn.: Wesleyan University Press, 2011), pp. 129–148.

25 WORD SIMPLY MEANT George Basalla, *Civilized Life in the Universe: Scientists on Intelligent Extraterrestrials* (Oxford University, 2006), pp. 56–62.

25 MEANDER LIKE STREAMS Flammarion, *Camille Flammarion's The Planet Mars,* pp. 373–382, 505–509.

25 ODDLY GEOMETRIC In 1882, Schiaparelli described the strange phenomenon of gemination, when one canal was suddenly joined by another running in parallel alongside it, like a set of railroad tracks. Schiaparelli generally believed bright areas were deserts, dark areas were seas, and half-toned areas were likely shallow seas or marshes. He also included the terms "island, isthmus, strait, channel, peninsula, cape, etc." In 1878, he underscored, "Our map . . . includes a complete system of geographical names which they who wish to avoid prejudice concerning the nature of the features on the planet may regard as a mere artifice

to assist the memory and abbreviate the descriptions. After all, we speak in a similar way of the *maria* of the Moon, knowing full well that they do not consist of liquid masses. If we understand the matter in this way, the names I have adopted will do no harm. . . ." William Sheehan and Stephen James O'Meara, *Mars: The Lure of the Red Planet* (Amherst, N.Y.: Prometheus Books, 2001), pp. 111–112; G. V. Schiaparelli, *Astronomical and Physical Observations of the Axis of Rotation and the Topography of the Planet Mars: First Memoir, 1877–1878* (San Francisco: Association of Lunar and Planetary Observers Monograph Number 5, 1994).

25 COMPLETED IN 1825 Jonathan Pearson, "Erie Canal Timeline," Union College (2003).

25 OPENED IN 1869 Charles Gordon Smith and William B. Fisher, "Suez Canal," *Encyclopedia Britannica* (updated Feb. 13, 2019).

25 WORK IN PANAMA Enrique Chaves, et al., "French Panama Canal Failure (1881–1889)," *The Panama Canal: A Triumph of American Medicine,* The University of Kansas Medical Center (March 13, 2019).

26 "HABITATION OF MARS" Flammarion, *Camille Flammarion's The Planet Mars,* pp. 373–382, 512.

26 "MAN OF MOODS" Louise Leonard, *Percival Lowell, An Afterglow* (Boston: Richard G. Badger, 1921), p. 15.

26 FASTEST POLO PONIES, Ibid., p. 29.

26 HERMIT AT HEART Ibid., pp. 19–20.

26 A CHRISTMAS GIFT William Sheehan, *The Planet Mars: A History of Observation and Discovery* (Tucson: University of Arizona Press, 1996), p. 104.

26 AN 1890 ARTICLE William H. Pickering, "Visual Observations of the Surface of Mars," *Sidereal Messenger,* 9 (1890), pp. 369–370.

27 ASTRONOMICAL VANTAGE POINTS Jordan D. Marché II, "Pickering, William Henry," in T. Hockley, et al., eds., *The Biographical Encyclopedia of Astronomers* (New York: Springer, 2007).

27 "SMOKE OF MEN" Percival Lowell, "Our Solar System," *Popular Astronomy,* 24 (1916), p. 419.

27 "BEST PROCURABLE AIR" Leonard, *Percival Lowell, An Afterglow,* p. 38. To select a location for the observatory, Lowell sent Andrew Ellicott Douglass, who had accompanied Pickering to Peru, ahead of him to Arizona in April of 1894. Armed with Lowell's fifteen-centimeter refractor, Douglass tested the seeing conditions at various sites. Since he only spent one or two days at any particular site, it was a completely unscientific study, and the decision to place the observatory on the mesa at Flagstaff was rather arbitrary. But Lowell was champing at the bit, and he decided that the altitude was an advantage (Douglass later said that Flagstaff also had the best saloons). In retrospect, sites in southern Arizona, especially around Tucson, would have been better—as Douglass later learned when he established the University of Arizona's Steward Observatory.

27 PICKERING DESIGNED Kevin S. Schindler, "100 Years of Good Seeing: The History of the 24-Inch Clark Telescope," Lowell Observatory (July 1996; revised September 1998), p. 1.

27 BY RAIL The fact that Flagstaff lay on a major rail line was another advantage in the siting of the facility.

27 GROUND WAS BROKEN Schindler, "100 Years of Good Seeing," p. 1.

27 FIRST OBSERVATIONS The first observations were made using borrowed telescopes, a forty-five-centimeter Brashear refractor and a thirty-centimeter Clark refractor.

27 SOON ACQUIRED The forty-five-centimeter Clark telescope arrived in July 1896.

27 A KITCHEN CHAIR Eric Betz, "Clark Telescope Going Dark," *Arizona Daily Sun* (Dec. 27, 2013).

27 "BUT ONE WATCHER" Leonard, *Percival Lowell, An Afterglow*, p. 27.

27 DOZENS MORE CANALS Percival Lowell, *Mars and Its Canals* (New York: The Macmillan Company, 1906).

27 "ALL-ENGROSSING MARTIAN PURSUIT" Percival Lowell, *Mars* (Boston: Houghton, Mifflin and Company, 1895), p. 128.

28 IN ENGLISH, FRENCH, AND GERMAN Leonard, *Percival Lowell, An Afterglow*, p. 27.

28 THE *BOSTON EVENING TRANSCRIPT* Ibid., pp. 25–27.

28 ON THE NEBULAR HYPOTHESIS Robert Markley, *Dying Planet: Mars in Science and the Imagination* (Durham, N.C.: Duke University Press, 2005), p. 66.

28 "SO SADLY TYPIFIED" Percival Lowell, "Mars (Part IV)," *The Atlantic* (Aug. 1895).

29 "THE OUTCOME IS DOUBTLESS YET" Percival Lowell, *Mars as the Abode of Life* (New York: The Macmillan Company, 1908), p. 135.

29 MOSTLY IN FOREIGN PERIODICALS As described in William H. Pickering, *Mars* (Boston: Richard G. Badger, 1921), p. 132. It should be noted that doubts were also expressed by some American astronomers, like Edward Emerson Barnard, who trained the great Lick ninety-one-centimeter refractor at Mount Hamilton, near San Jose, California, on Mars at the same moment that Lowell was cutting his teeth with the forty-five-centimeter refractor at Flagstaff (and before Lowell had published his theory of intelligent life on the planet). "I have been watching and drawing the surface of Mars," Barnard wrote to a colleague. "To save my soul I can't believe in the canals as Schiaparelli draws them. . . . I verily believe—for all the verifications—that the canals . . . are a fallacy and that they will be so proved before many oppositions are past." William Sheehan, *The Immortal Fire Within: The Life and Work of Edward Emerson Barnard* (Cambridge University Press, 1995), p. 246.

29 A BRITISH SOLAR ASTRONOMER The astronomer was Edward Walter Maunder, who worked at the Royal Observatory in Greenwich.

29 TO CHALLENGE LOWELL Maunder was joined in his skepticism of Lowell by Barnard as well as by Vincenzo Cerulli, who operated a private observatory equipped with a thirty-nine-centimeter refractor near Teramo, Italy. Cerulli had sketched canals with the best of them until, on January 4, 1897, he noticed, in moments of "perfect definition [in which] Mars appeared perfectly free from

undulation," that one of the canals, the Lethes, suddenly "lost its form of a line and altered itself into a complex and indecipherable system of minute patches." Henceforth he argued that the canals were mere tricks of the eye, formed when small irregular detached details were visualized in unsteady air or were present at the threshold of resolvability. Sheehan, *The Planet Mars,* p. 125.

29 A GROUP OF SCHOOLBOYS J. E. Evans and E. W. Maunder, "Experiments as to the Actuality of the 'Canals' Observed on Mars," *Monthly Notices of the Royal Astronomical Society,* 63 (1903), pp. 488–499.

29 COUNTERED THAT LINEAR FEATURES Lowell was also unimpressed with the testimony of the students in a "reform school."

29 TAKEN ABACK The most prominent early member of Lowell's staff, Andrew Ellicott Douglass, who had been with the observatory from its founding, had also turned skeptical; Douglass was dismissed in 1901 but landed on his feet, going on to found the astronomy department at the University of Arizona in Tucson.

29 LOWELL TURNED TO PHOTOGRAPHY K. Maria D. Lane, "Mapping the Mars Canal Mania: Cartographic Projection and the Creation of a Popular Icon," *Imago Mundi,* 58: 2 (2006), pp. 198–211.

29 REVEALED NEW MOONS Following the discovery of Phoebe by William Pickering in 1899, Himalia and Elara, two new moons of Jupiter, were discovered by photography in 1904 and 1905.

29 TOOK ON THE CHALLENGE The assistant was Carl O. Lampland, who designed several astronomical cameras. Instead of traveling to the Andes, he remained in Flagstaff for the 1907 opposition to conduct similar imaging of Mars.

29 DISTRIBUTED THEM WIDELY Schiaparelli, when informed of the accomplishment, excitedly wrote to Lowell, "I should never have believed it possible."

29 FOLLOWED SUIT IN 1906 At the meeting in June of that year, President A.C.D. Crommelin went on to convey that "it seemed to him clear that the photographs could hardly be an illusion as to the existence of some approximately linear markings there." See: Lane, "Mapping the Mars Canal Mania," p. 205; "Report of the Meeting of the Association, Held on June 20, 1906, at Sion College, Victoria Embankment," *Journal of the British Astronomical Association,* 16, no. 9 (1906), p. 333.

30 EVEN BETTER PHOTOGRAPHS Lane, "Mapping the Mars Canal Mania," pp. 198–211; Simon Newcomb, "The Optical and Psychological Principles Involved in the Interpretation of the So-Called Canals of Mars," *Astrophysical Journal,* 26:1 (1907), pp. 1–17.

30 THE MUCH-HYPED EXPEDITION WAS David Peck Todd, "The Lowell Expedition to the Andes," *Popular Astronomy,* 15 (1907), pp. 551–553; William Sheehan and Anthony Misch, "The Great Mars Chase of 1907," *Sky & Telescope* (November 2007), pp. 20–24.

30 IN THE OPEN AIR No dome was needed, for it hardly ever rained.

30 SOME SEVENTY KILOMETERS INLAND Hilmar W. Duerbeck, "National and International Astronomical Activities in Chile 1849–2002." In *Interplay of Periodic,*

Cyclic and Stochastic Variability in Selected Areas of the H-R Diagram, 292 (2003), pp. 3–20.

30 THE CENTURY MAGAZINE At the end of the expedition, Lowell and Todd descended into mutual vituperations as each claimed to have publication rights to details of the expedition. Legal action was threatened. In the end, Todd published an article in *Cosmopolitan,* but Lowell won exclusive rights to publication of the images. See: Percival Lowell, "New Photographs of Mars: Taken by the Astronomical Expedition to the Andes and Now First Published," *The Century Magazine,* 75 (1907), pp. 303–311; E. C. Slipher, "Photographing Mars," *The Century Magazine,* 75 (1907), p. 312; K. Maria D. Lane, *Geographies of Mars: Seeing and Knowing the Red Planet.* (Chicago: University of Chicago Press, 2011), pp. 118–120.

30 THE DISCLAIMER Lane, "Mapping the Mars Canal Mania," pp. 198–211.

30 "[THE CANALS] ARE THERE" Lowell, "New photographs of Mars," *The Century Magazine,* pp. 303–311.

30 LAUNCHED AN ATTACK Alfred Russel Wallace, *Is Mars Habitable? A Critical Examination of Professor Percival Lowell's Book "Mars and Its Canals," with an Alternate Explanation* (London: Macmillan and Co., Ltd., 1907), pp. 55–77.

30 THE GRECO-FRENCH ASTRONOMER EUGÈNE ANTONIADI Antoniadi was an exceptionally skillful artist as well as an astronomer, who had worked since the 1890s with Flammarion at the latter's observatory at Juvisy (near Paris). He fell out with Flammarion—both were very strong personalities—and married a Greek woman with independent means in 1902. He pursued interests other than astronomy for several years, including the production of a three-volume study, in Greek, of the Hagia Sophia in Istanbul, through which he honed his artistic skills. He also took up chess with a passion and eventually became a near grand master. Using the "Grande Lunette" at the Meudon Observatory, near Paris (then and still the largest refractor in Europe), Antoniadi saw not canals but "a vast and incredible amount of detail held steadily, all natural and logical, irregular and chequered, from which geometry was conspicuous by its complete absence." Lowell, of course, did not accept this verdict and argued that Antoniadi, whom he called "a man without knowledge of how to observe," had been tricked by atmospheric effects, which had blurred the actual lines present on the surface, making them appear irregular and discontinuous. Sheehan and O'Meara, *Mars: The Lure of the Red Planet,* pp. 155–181.

31 "NATURAL AGENCIES OF" Lane, "Mapping the Mars Canal Mania," pp. 198–211; E. M. Antoniadi, "On the Possibility of Explaining on a Geomorphic Basis the Phenomena Presented by the Planet Mars," *Journal of the British Astronomical Association,* 20:2 (1909), p. 93.

31 THE PIONEERING PSYCHOLOGISTS For more on the psychology of planetary perception, see Chapter 14, "A Stately Pleasure Dome," in William Sheehan, *Planets and Perception: Telescopic Views and Interpretations, 1609–1909* (University of Arizona Press, 1988).

31 THEORY OF SPECIAL RELATIVITY Albert Einstein, "Zur Elektrodynamik bewegter Körper," *Annalen der Physik,* 322, no. 10 (1905), pp. 891–921.

31 TO A BACKWATER Much of the useful work done during the interwar years was done by amateurs.

31 "WARMTH OF HIS FIRE" Leonard, *Percival Lowell, An Afterglow,* p. 42.

31 "'LIGHT THAT SHIFTS'" This is a line from "To the True Romance," a poem by Rudyard Kipling.

31 SOUTH AFRICA Those expeditions, sponsored by the National Geographic Society to take advantage of the planet's higher elevation above the horizon, took place in 1939 and 1954.

31 ONE HUNDRED THOUSAND IMAGES William Sheehan, *The Planet Mars,* p. 146.

31 IN 1962 The last opposition Slipher photographed was a year later, in 1963. He died in 1964, a few months before *Mariner 4* set out for the Red Planet.

32 "A VAST COLLECTION" Earl Slipher, *The Photographic Story of Mars* (Cambridge, Mass.: Sky Pub. Corp., 1962).

32 "SO-CALLED 'CANALS'" Peter M. Millman, *This Universe of Space* (Toronto: Canadian Broadcasting Corporation, 1961), pp. 26, 28.

32 *THE BOOK OF MARS* Samuel Glasstone, *The Book of Mars* (Washington, D.C.: Scientific and Technical Information Division, Office of Technology Utilization, NASA, 1968), p. 126.

32 STRUGGLED TO CORRELATE One planetary scientist remarked that the *Mariner 4* images were like staring through binoculars at the wrinkled skin of an elephant. *Mariner 6* and *7* were designed to cover this "missing middle," helping to bridge the gap and allow for a reliable interpretation of the close-up images of the Martian surface. Stewart A. Collins, *The Mariner 6 and 7 Pictures of Mars* (Pasadena, Calif.: NASA JPL, 1971), pp. 24–25.

32 NASA PRESS TEAM "Press Kit, Mariner Mars '69," NASA (Feb. 14, 1969).

33 BEGAN A ROUTINE TEST PROCEDURE Kay Grinter, "One small step on the Moon, one giant footprint on Mars," *Spaceport News* (March 26, 2004); DNews, "The Brave Story of Mars' McClure-Beverlin Escarpment," *Seeker* (March 3, 2014); James H. Wilson, "Two over Mars—Mariner 6 and Mariner 7, February–August 1969" (1970), p. 13; John Casani, personal interview by Sarah Johnson (Pasadena, Calif., Aug. 6, 2015).

33 TWO MEMBERS The members of the crew were Bill McClure and Jack Beverlin; in 2014, the McClure-Beverlin Escarpment on Mars was informally named in their honor.

34 THE TELEPHOTO SHUTTERS Collins, *The Mariner 6 and 7 Pictures of Mars,* p. 24.

34 SEEMED EVIDENT THAT COPRATES In *Mariner 7* frames 7F69 and 7F70, the classical feature Coprates, named after an ancient Persian river, appeared as a collection of dark dots, leading to the conclusion that "the 'canal' Coprates can now be identified as a sequence of separate dark features." See: Collins, *The Mariner 6 and 7 Pictures of Mars,* p. 58. Yet a different view emerged with the much higher resolution imaging from *Mariner 9*. In fact, the newly discovered Coprates Chasma, a canyon structure and part of Valles Marineris, was continuous and relatively linear. "With the exception of the extraordinary canyon that appears to coincide with the rather stubby 'canal' Coprates, no other clear-cut features have been found to account for the system of canals reported by many observers." See: William K. Hartmann and Odell Raper, *The New Mars: The Dis-*

coveries of Mariner 9 (Washington, D.C.: NASA Office of Space Science, 1974), p. 63.

34 A "LUMP" Collins, *The Mariner 6 and 7 Pictures of Mars,* p. 65.

34 W-SHAPED CLOUDS Ibid., p.59.

35 CRISP FALL DAYS Ibid., p. 20.

35 20 PERCENT OF THE MARTIAN SURFACE Ibid., p. 24.

35 ESTABLISH THE MARTIAN GEOID "Press Kit, Project: Mariner 9," NASA (Oct. 22, 1971).

35 *MARINER 8* WOULD MAP Norman Haynes, personal interview by Sarah Johnson (Pasadena, Calif., Aug. 6, 2016).

36 BLASTED OFF ON MAY 8 The launch date was 01:11:02 UTC on May 9 (late on May 8 in the United States); John Noble Wilford, "Mariner 8's Rocket Fails After Lift-off, Dooming Mars Trip," *The New York Times* (May 9, 1971).

36 MAROONING THE SPACECRAFT "Kosmos 419," NASA Science Solar System Exploration (Jan. 26, 2018).

36 EIGHT-DIGIT CODE Asif A. Siddiqi, *Deep Space Chronicle: A Chronology of Deep Space and Planetary Probes 1958–2000,* Monographs in Aerospace History, no. 24 (2017) p. 86.

36 SUNFLOWER SEED Simulations indicated that it was "an integrated circuit chip one twentieth of an inch square" that failed; "Mariner I Assigned New Mission," NASA JPL (May 26, 1971).

36 A FAULTY DIODE Ibid.

Chapter 3: Red Smoke

37 AMONG THE FIRST Charles F. Capen and Leonard J. Martin, "The Developing Stages of the Martian Yellow Storm of 1971," *Lowell Observatory Bulletin,* no. 157 (Nov. 30, 1971), p. 211. At the much-anticipated opposition of Mars of 1956, when the planet came within 56 million kilometers of Earth, a "Great Dust Storm" encircled the planet. Though dust clouds had been common, the scale of this dust storm took astronomers by surprise. Generally, it was regarded as rather anomalous, though during early 1971, Chick Capen, an astronomer at the Lowell Observatory's Planetary Research Center, predicted that another such event was likely to occur that year—because in 1971, Mars's closest approach to Earth would happen very near to the time of its perihelion passage, when it was most strongly heated by the sun. By late September and early October, amateur and professional astronomers around the world were observing the dust clouds.

37 SMOOTH, LACQUERED CLOUD Capen and Martin, "The Developing Stages of the Martian Yellow Storm of 1971," p. 214.

37 "A BILLIARD BALL" Norman Haynes, personal interview by Sarah Johnson (Pasadena, Calif., Aug. 6, 2016).

38 REPROGRAMMED THE COMPUTER SYSTEM JPL installed a computer with reprogrammable memory on the *Mariner 6* and *7* missions as well. It was tested in flight, but it wasn't used for anything critical during the mission. The *Mariner 9*

spacecraft employed a similar design to the *Mariner 6* and *7* spacecraft design but with a modified tape recorder with higher capacity. The reprogrammable memory on *Mariner 6* and *7* was 128 words; for *Mariner 9,* it was expanded to 512 words, enough to program the sequence of science observations (varying the orbital geometry and viewing conditions) for each orbit.

38 IMPENETRABLE DUST CLOUDS V. G. Perminov, "The Difficult Road to Mars: A Brief History of Mars Exploration in the Soviet Union" (Washington, D.C.: Monographs in Aerospace History, no. 15, 1999), p. 59. The Soviets didn't have the same luxury of waiting out the storm, as their software was not capable of remote reprogramming.

38 SMALL TETHERED ROBOT Amy Shira Teitel, "The Soviet Rovers That Died on Mars," *Discover* (July 20, 2017).

38 SUNLIGHT WARMED THE SURFACE Caleb A. Scharf, "The Great Martian Storm of '71," *Scientific American* (Oct. 21, 2013).

39 JSC-MARS-1A Carlton C. Allen, et al., "JSC-Mars-1: Martian Regolith Simulant," Lunar and Planetary Science Conference, 28 (1997).

39 IN THE SADDLE Inge Loes ten Kate, "Organics on Mars Laboratory Studies of Organic Material Under Simulated Martian Conditions," Doctoral thesis, Leiden University (2006) p. 76.

40 AS FINE AS TALCUM POWDER Atmospheric dust particles have been estimated to be approximately 3 μm in diameter. M. T. Lemmon, et al., "Atmospheric Imaging Results from the Mars Exploration Rovers: Spirit and Opportunity," *Science,* 306, no. 5,702 (2004), p. 1,753.

41 "WAVE OF DARKENING" Sheehan and O'Meara, *Mars: The Lure of the Red Planet*, p. 354; Caleb A. Scharf, "Mars and the Wave of Darkening," *Scientific American* (Aug. 9, 2018).

41 "TOUCH OF MOSS GREEN" Gerard P. Kuiper, "Visual Observations of Mars, 1956," *The Astrophysical Journal,* 125 (1957), p. 307. Even though none of the diagnostic spectral features of chlorophyll were seen, a Soviet scientist named Gavriil Tikhov then showed that chlorophyll-absorption bands can extend and even disappear in tundra conditions, particularly where oxygen is limited.

41 "THIS EVIDENCE" William M. Sinton, "Spectroscopic Evidence for Vegetation on Mars," *Astrophysical Journal,* 126 (1957), p. 231; Sinton, "Further Evidence of Vegetation on Mars," *Science,* 130, no. 3,384 (1959), pp. 1,234–1,237; Steven J. Dick, *Life on Other Worlds: The 20th-Century Extraterrestrial Life Debate* (Cambridge University Press, 2001), p. 51.

41 BY 1962 The French colleague was Jean-Henri Focas. "Observations of Mars Made in 1961 at the Pic Du Midi Observatory," NASA Technical Report, JPL-TR-32-151 (1962).

42 HAD BEEN IMPOSSIBLE William K. Hartmann and Odell Raper, *The New Mars: The Discoveries of Mariner 9* (Washington, D.C.: NASA Office of Space Science, 1974), p. 17.

42 WERE EVIDENCE OF LIFE "Press Kit: Project: Mariner Mars 1971," NASA (April 30, 1971).

42 WILLIAM PICKERING E. P. Martz, Jr., "Professor William Henry Pickering, 1858–1938, An Appreciation," *Popular Astronomy*, 46, no. 456 (June–July 1938), p. 299; Leon Campbell, "William Henry Pickering, 1858–1938," *Publications of the Astronomical Society of the Pacific*, 50, no. 294 (1938), pp. 122–125.

42 FIRST RECREATIONAL GUIDE William Henry Pickering, *Guide to the Mt. Washington Range* (Boston: A. Williams, 1882).

42 AND ADVICE ABOUT Ibid., p. 10.

42 "ABOVE THE TREE-LINE" Ibid., p. 11.

42 PREFERRED WILD PLACES As an astronomer, Pickering was very involved in scouting favorable locations in which to set up observatories. In 1889, he was the first person to test the suitability of Mount Wilson in California for astronomical observations. In the decades that followed, Mount Wilson Observatory became one of the most famous observatories in the world and the birthplace of modern observational cosmology (see "Our Story," Mount Wilson Observatory, www.mtwilson.edu).

42 "DARWIN'S BULLDOG" Paul White, *Thomas Huxley: Making the "Man of Science"* (Cambridge University Press, 2003).

43 "GOOD SEEING" William H. Pickering, *Mars* (Boston: Richard G. Badger, 1921), p. 132.

43 ELECTRODYNAMICS OR PHYSIOLOGY Kristina Maria Doyle Lane, "Imaginative Geographies of Mars: The Science and Significance of the Red Planet, 1877–1910," doctoral thesis, University of Texas at Austin (2006) p. 90; William H. Pickering, "The Planet Mars," *Technical World Magazine* (1906), pp. 463–464.

43 MANY ALTERNATIVE THEORIES Pickering had lots of ideas about everything. In contrast to Lowell, who once his mind was made up about something stuck to it without wavering, Pickering seemed to have been only lightly attached to any of his own ideas. For more on his theories of life on Mars, see William H. Pickering, "Report on Mars, No. 37: What I Believe About Mars," *Popular Astronomy*, 34 (1926), pp. 482–491.

43 WERE *THEMSELVES* VEGETATION David Bressan, "The Earth-like Mars," *Scientific American*, 14 (Aug. 2012).

43 NATURALLY FORMED CRACKS Pickering, "Report on Mars, No. 37," *Popular Astronomy*, pp. 482–491.

43 PASTURES FOR CATTLE Pickering, *Mars*, pp. 149–150.

43 "MORE OR LESS CONTINUOUS" Ibid., p. 150.

44 ONE-STORY PLANTATION Howard Plotkin, "William H. Pickering in Jamaica: The Founding of Woodlawn and Studies of Mars," *Journal for the History of Astronomy*, xxiv (1993), p. 109; Philip M. Sadler, "William Pickering's Search for a Planet Beyond Neptune," *Journal for the History of Astronomy*, 21, no. 1 (Feb. 1990), pp. 59–60.

44 "WHOOP ABOUT THE STARS" Plotkin, "William H. Pickering in Jamaica," *Journal for the History of Astronomy*, p. 111.

44 "THE ENORMOUS SIZE" William H. Pickering, "Island Universes and the Origin of the Solar System," *The Observatory*, 47 (1924), p. 56.

44 "HAD IT NOT BEEN SETTLED" Pickering, *Mars*, pp. 156–157.

45 CABLE HIS REPORTS Sadler, "William Pickering's Search for a Planet Beyond Neptune," *Journal for the History of Astronomy*, p. 60; E. P. Martz, Jr., "Pilgrimage to a Tropical Observatory," *Popular Astronomy*, 45 (1937), pp. 419–428.

45 "NOT FORTUNATE ENOUGH" William H. Pickering, "Monthly Report on Mars—No. 1," *Popular Astronomy*, 22 (1914), p. 1.

45 FOR YEARS In total, there were forty-four reports, published between 1913 and 1930. Martz, "Professor William Henry Pickering," *Popular Astronomy*, p. 301.

45 PENCIL SKETCHES Pickering, "Instrument Readings, Notes, and Landscape Sketches, 1891–1892," *Papers of William Henry Pickering, 1870–1907* (Harvard University Archives, HUG 1691, HUG 1691.65).

45 SHADES OF SIENNA Pickering, *Mars*, p. 28.

45 "MUST REIGN SUPREME" Pickering, quoted in Sadler, "William Pickering's Search for a Planet Beyond Neptune," *Journal for the History of Astronomy*, p. 60.

45 "IN ABEYANCE" Pickering, "Monthly Report on Mars—No. 4," *Popular Astronomy*, 22 (1914), p. 228.

45 BLUE-TINTED BAYS Pickering, "Monthly Report on Mars—No. 2," *Popular Astronomy*, 22 (1914), p. 96.

45 GREENING OF THE SOUTHERN *MARIA* Ibid., p. 94.

45 TORRENTS OF SIBERIA Pickering, "Monthly Report on Mars," *Popular Astronomy*, 22 (1914), pp. 3–4.

45 ABOVE WESTERN BOLIVIA Ibid., p. 4.

45 STIPPLED WITH HOARFROST Pickering, "Monthly Report on Mars—No. 4," *Popular Astronomy*, p. 224.

45 "SNOWED UNDER" Pickering, "Monthly Report on Mars.—No. 2," *Popular Astronomy*, p. 92.

46 DEPTH OF SEVERAL METERS Ibid., p. 99.

46 WITH VACUUM THERMOCOUPLES While the original experiments measured planetary radiation as percentages of the total radiation, some researchers converted those results into thermometric degrees, suggesting more accuracy than justified. See: Steven J. Dick, *Life on Other Worlds: The 20th-Century Extraterrestrial Life Debate* (Cambridge University Press, 2001), pp. 45–47; W. W. Coblentz, "Thermocouple Measurements of Stellar and Planetary Radiation," *Popular Astronomy*, 31 (1923), pp. 105–121.

46 "GRANDEUR AND LONELINESS" Pickering, *Guide to the Mt. Washington Range*, p. 11.

46 A COMMONLY HELD BELIEF In part, this was due to the fact that observers failed to detect any irregularities and notches along the terminator on Mars, such as are visible on the moon (the presence of dust in the atmosphere largely explains this smoothness).

46 FINAL VEGETATION THEORY It should be noted that Pickering maintained an open mind as to higher life-forms. In his thirty-seventh report, Pickering wrote: "The fourth explanation for the canals is the one to which I definitely adhere for the transfer of water from pole to pole, and back again. It involves no artificial aid whatsoever, but I am nevertheless far from denying the possibility of the existence of animal life, and even intelligent animal life upon our neighboring planet." Pickering, "Monthly Report on Mars—No. 37: What I Believe About Mars," *Popular Astronomy*, 34 (1926), p. 484.

46 APPEARED TO BE RECEDING "Mariner 9," NASA Science: Solar System Exploration (July 31, 2019).

47 MORE SPOTS SLOWLY APPEARED Haynes, personal interview by Johnson.

47 HEAD OF THE IMAGING TEAM BLURTED Ibid.

47 MOUNTAINOUS VOLCANOES Bruce Murray, interviewed by Rachel Prud'homme, audiocassette recording transcript, California Institute of Technology Archives and Special Collections (Pasadena, Calif., 1993), p. 82.

47 ONE OF THE LARGEST The central peak of Rheasilva on the asteroid Vesta is slightly taller than Olympus Mons, though the diameter of Olympus Mons is greater than Vesta itself.

47 "WE SAW THEM COMING" Haynes, personal interview by Johnson.

48 EARLY STUDIES HAD NOTED William Sheehan, *The Planet Mars: A History of Observation and Discovery* (Tucson: University of Arizona Press, 1999), p. 156.

48 ALMOST IMPOSSIBLE TO BELIEVE Hartmann and Raper, *The New Mars: The Discoveries of Mariner 9*, p. 94.

48 MANY ON THE TEAM WONDERED, Ibid., p. 97.

49 EVEN A KNEE-DEEP RIVULET. Ibid.

Chapter 4: The Gates of the Wonder World

53 A PUBLIC-BROADCASTING RECORD Bill Carter, "'Civil War' Sets an Audience Record for PBS," *The New York Times* (Sept. 25, 1990).

53 "SNOWBALLS OF SATURN" Carl Sagan, Ann Druyan, and Steven Soter, writers, *Cosmos*, Season 1, episode 6, "Travellers' Tales," directed by Adrian Malone, et al. Aired Nov. 2, 1980. Australian Broadcasting Commission, Carl Sagan Productions, and KCET, 1980.

54 HALF A BILLION VIEWERS David A. Hollinger, "Star Power: Two Biographies of Carl Sagan Explore the Scientist as Celebrity and the Celebrity as Scientist," *The New York Times* (Nov. 28, 1999).

54 SMOKED LOTS OF MARIJUANA Keay Davidson, *Carl Sagan: A Life* (New York: John Wiley & Sons, 1999), p. 214. Biographies containing stories of Sagan's early life and career, from which many of the details in this chapter are drawn, include Spangenburg, Moser, and Moser, *Carl Sagan: A Biography*; William Poundstone, *Carl Sagan: A Life in the Cosmos* (New York: Henry Holt, 1999); and Keay Davidson, *Carl Sagan: A Life*.

54 TURTLE-LIKE CREATURES Carl Sagan, *Carl Sagan's Cosmic Connection: An Extraterrestrial Perspective* (Cambridge University Press, 2000), p. 45.

54 "LARGE ORGANISMS . . ." Carl Sagan and Joshua Lederberg, "The Prospects for Life on Mars: A Pre-Viking Assessment," *Icarus,* 28 (1976), p. 291.

54 THESE CREATURES, HE SPECULATED Ibid., p. 297; George Basalla, *Civilized Life in the Universe: Scientists on Intelligent Extraterrestrials* (Oxford University Press, 2006), p. 110; "Mars: The Search Begins," *Time,* 108 (July 5, 1976), pp. 87–90; Carl Sagan, *Other Worlds* (New York: Bantam Books, 1975).

54 DRINKING HYDRATED MINERALS Sagan and Lederberg, "The Prospects for Life on Mars," *Icarus,* pp. 295–296.

55 SILICON-BASED GIRAFFES Sagan, "The Search for Extraterrestrial Life," *Scientific American,* 271, 4 (October 1994), p. 93.

55 "SHOW UP AS A STREAK" David S. Salisbury, "Will Viking Find Life on Mars?" *The Lowell Sun* (July 8, 1976).

55 INTO A NEARBY FIELD Carl Sagan, *Cosmos* (New York: Ballantine Books, 1985), pp. 90–91.

55 BLOCK-LETTER HEADLINES Sean Hutchinson, "15 Highlights from Carl Sagan's Archive," *Mental Floss* (Feb. 6, 2014).

55 SHORT TECHNICAL VOLUME Arthur C. Clarke, *Interplanetary Flight: An Introduction to Aeronautics* (New York: Harper, 1952).

55 THE MAYO CLINIC Ray Spangenburg, Kit Moser, and Diane Moser, *Carl Sagan: A Biography* (Westport, Conn.: Greenwood Publishing Company, 2004), p. 12.

56 DIAGNOSED WITH ACHALASIA Ibid.

56 OBSESSIVE, NEUROTIC INFLUENCE Keay Davidson, *Carl Sagan: A Life* (New York: John Wiley & Sons, 1999), pp. 2, 9–11, 42.

56 FILLED WITH BLOOD Jorge Alberto Delucca, *A Few Great Scientists: From Alfred Nobel to Carl Sagan* (Bloomington, Ind.: Xlibris Corporation, 2017).

56 THESIS ON THE ORIGINS OF LIFE Poundstone, *Carl Sagan: A Life in the Cosmos,* p. 25.

56 PHD RESEARCH Carl Sagan, *Physical Studies of Planets* (University of Chicago PhD thesis, 1960).

56 HELP THE NATIONAL ACADEMY OF SCIENCES In 1958, the National Academy of Sciences decided to more closely examine the idea of life beyond the Earth. Joshua Lederberg, a Nobel Prize–winning geneticist at the University of Wisconsin, was invited to co-chair a panel on extraterrestrial life. After the group decided to draft a handbook of planetary biology, Lederberg brought up the possibility of a NASA contract for young Sagan. He wrote: "This really is a substantial job, and there is some problem in finding a sufficiently informed enthusiast to do the work. Fortunately Mr. Carl Sagan may be available for some months this summer, and perhaps again after he completes his dissertation in astronomy (planetary atmospheres) at the Yerkes Observatory." Memorandum from R. C. Peavey to the Space Science Board, the Committee on Space Projects, and the Committee on Psychological and Biological Research, 13 April

1959, Joshua Lederberg Papers, 1904–2008. Located in: Archives and Modern Manuscripts Collection, History of Medicine Division, National Library of Medicine, Bethesda, Md.; MS C 552. For a fantastic graphic history of these events, see also: Mary Voytek, et al., *Astrobiology: The Story of Our Search for Life in the Universe* (Mountain View, Calif.: NASA Astrobiology Program, 2010).

56 "SPADEWORK, MAINLY CONSULTATION" E. C. Levinthal, Cytochemical Studies of Planetary Microorganisms Explorations in Exobiology, NASA Technical Report No. IRL1213 (Washington: NASA, 1980), Attachment 1: "March 4, 1959 Letter from Lederberg to Jastrow." See also: James E. Strick, "Creating a Cosmic Discipline: the Crystallization and Consolidation of Exobiology, 1957–1973," *Journal of the History of Biology* 37, no. 1 (2004), pp. 131–180.

56 THINGS LIKE STERILIZATION Jacob Berkowitz, *The Stardust Revolution: The New Story of Our Origin in the Stars* (Buffalo, N.Y.: Prometheus Books, 2012), p. 132.

56 RISKING HUMANITY'S CHANCE "[S]ince the sending of rockets to crash on the moon's surface is within the grasp of present technique, while the retrieval of samples is not," Lederberg had argued in a manuscript he put together for *Science,* "we are in the awkward situation of being able to spoil certain possibilities for scientific investigation for a considerable interval before we can constructively realize them." Joshua Lederberg and Dean B. Cowie, "Moondust," *Science,* vol. 127, no. 3313 (1958), pp. 1,473–1,475.

57 NATIONAL ACADEMIES PANEL COALESCED WESTEX and EASTEX quickly commenced regular meetings, chaired respectively by Lederberg and Melvin Calvin, a chemistry professor at Berkeley.

57 AN EAST COAST GROUP, EASTEX, AND Steven J. Dick and James E. Strick, *The Living Universe: NASA and the Development of Astrobiology* (New Brunswick, N.J.: Rutgers University Press, 2004), p. 25.

57 BOMBASTICALLY LAMENTED That member was Thomas Gold, a Cornell astronomy professor. The moment is described in Carl Sagan, "Wolf Vladimir Vishniac: An Obituary," *Icarus,* vol. 22, issue 3 (1974), pp. 397–398.

57 "UNDOUBTEDLY THE WORLD'S" Eugene Kinkead, "The Tiny Landscape, Pt. 1," *The New Yorker* (July 2, 1955), p. 29.

57 "NATURE, GOD, WHATEVER" Ibid.

57 FED TO SEAHORSES Kinkead, "The Tiny Landscape, Pt. 2," *The New Yorker* (July 9, 1955), p. 39.

57 AMERICAN EXPORT LINES STEAMER Maya Benton, *Roman Vishniac Rediscovered* (New York: Prestel, 2015); Roman Vishniac, "Wolf Vishniac arriving with his family in New York Harbor on the S.S. Siboney, New York," ca. 1940 (New York: International Center of Photography, 2013).

57 SOME DETAILS OF PHOTOSYNTHESIS Wolf Vishniac, Bo L. Horecker, and Severo Ochoa, "Enzymic Aspects of Photosynthesis," *Advances in Enzymology and Related Areas of Molecular Biology,* 19 (1957), pp. 1–77.

57 HOW MICROBES USED SULFUR Wolf Vishniac and Melvin Santer, "The Thiobacilli," *Bacteriological Reviews,* 21, no. 3 (1957), p. 195.

58 BUTTERFINGERED Poundstone, *Carl Sagan: A Life in the Cosmos,* p. 49.

58 VISHNIAC HAD A QUIET NATURE Wolf Vishniac, "Letter to Senator Clinton P. Anderson," U.S. National Library of Medicine (Aug. 28, 1969).

58 IN 1959 "The Search for Martian Life Begins: 1959–1965," in Edward Clinton Ezell and Linda Neuman Ezell, *On Mars: Exploration of the Red Planet, 1958–1978* (Washington, D.C.: The NASA History Series, 1984).

58 "WOLF TRAP" Wolf Vishniac, "Extraterrestrial Microbiology," *Aerospace Medicine* (1960), pp. 678–680; "The Search for Martian Life Begins: 1959–1965," in Ezell and Ezell, *On Mars: Exploration of the Red Planet, 1958–1978.*

58 WORKING MODEL Vishniac originally developed the device to demonstrate "the feasibility of automatic remote detection of the growth of microorganisms . . . [h]e wanted to prove that such an instrument could be built." In 1961, he arranged a contract with Ball Brothers Research Corporation to develop a more complex breadboard. "The Search for Martian Life Begins: 1959–1965," in Ezell and Ezell, *On Mars: Exploration of the Red Planet, 1958–1978.*

58 ONE OF THE MOST INFLUENTIAL PALEONTOLOGISTS The elder Simpson was George Gaylord Simpson, who served as the curator of Harvard's Museum of Comparative Zoology from 1959 until 1970.

59 TEASED THEM ABOUT George Gaylord Simpson, "The Nonprevalence of Humanoids," in *This View of Life: The World of an Evolutionist* (New York: Harcourt, Brace & World, 1964), pp. 253–254; John D. Rummel, "Carl Woese, Dick Young, and the Roots of Astrobiology," *RNA Biology,* 11, no. 3 (2014), pp. 207–209.

59 INSTRUMENT DUBBED "GULLIVER" David Warmflash, "Celebrating Viking: Gilbert Levin Recalls the Search for Life on Mars," *Discover* (July 20, 2016).

59 GULLIVER SOUGHT TO CAPITALIZE Gil Levin, the instrument's inventor, was a sanitation engineer who worked in public-health engineering. The typical test was based on a simple idea: If many bacteria were present in a water sample, a lot of carbon dioxide would be respired after the sample was added to nutrient-rich media, but if only a little carbon dioxide was respired, below a certain threshold, the water was considered safe. Yet the typical method took a week or longer, which irked Levin, as no one was drinking last week's water or swimming in last week's ocean. Levin had been working away on a new concept, which would utilize radioactivity to make the same test thousands of times more sensitive. The idea was to monitor the release of carbon dioxide using a radioactive form of carbon. If organisms were present, and if they consumed nutrients made from carbon-14, the carbon-14 could be respired back into the air as carbon dioxide and registered on an inexpensive but highly sensitive Geiger counter. Only a tiny whiff of the gas would need to be present, which could massively expedite the detection time. Jay Gallentine, "What If," *Infinity Beckoned: Adventuring through the Solar System, 1969–1989* (Lincoln: University of Nebraska Press, 2016).

59 KITE LINE Once the kite line was sealed in a growth chamber, an ampule of labeled organic nutrients would be broken. If the nutrients were metabolized, the exhaled breath of the microbes would be trapped by a chemically coated film at the mouth of a shockproof Geiger counter.

59 NEARLY TWENTY Gallentine, *Infinity Beckoned: Adventuring through the Solar System, 1969–1989,* p. 17.

59 REQUIRED TOO MUCH DATA The idea of microscopic imaging had been considered, but the transmission of a vidicon image would likely require "10^5 bits for a bad picture"—the equivalent of a text message today—or "10^7 bits for a good picture"—about the size of a photo snapped by the first model of the iPhone. "When this very large data requirement and the problems of specimen preparation and slide searching were considered, no further attention was given to this device." Life Detection Experiments Team, "A Survey of Life Detection Instruments for Mars," NASA TMX-54946 Technical Report, NASA (August 1963), p. 15.

59 SOMETHING YOU MEASURED The prototypes of life detection instruments focused almost entirely on microbes: Microbes were always associated with higher life-forms. Microbes were fecund, and microbes were everywhere. It would be hard to find a bucket of soil anywhere on the Earth that wasn't teeming with them. Detecting microbes could also be done at the micro-scale, an advantage with the space and weight for hardware so highly constrained.

60 PENNED A MOVING LETTER Vishniac, "Letter from Wolf Vishniac to Clinton P. Anderson, United States Senate."

60 SCRUTINIZED HUNDREDS OF PHOTOGRAPHS S. D. Kilston, R. R. Drummond, and C. Sagan, "A Search for Life on Earth at Kilometer Resolution," *Icarus*, 5, 79 (1966), pp. 79–98; Carl Sagan and David Wallace, "A Search for Life on Earth at 100 Meter Resolution," *Icarus*, 15, 3 (1970), pp. 515–554; Carl Sagan, "Is There Life on Earth?" *Engineering and Science*, 35 (4), (1972), pp. 16–19.

60 NECESSARILY PRECLUDE LIFE Carl Sagan, "Statement of Dr. Carl Sagan, Department of Astronomy, Cornell University, Ithaca, N.Y.," Symposium on Unidentified Flying Objects, Hearings Before the Committee on Science and Aeronautics, U.S. House of Representatives, 90th Congress, 2nd Session (July 29, 1968).

60 LITTLE-KNOWN EXPERIMENTS The original "Mars jars" work began in the lab of military space medicine expert Hubertus Strughold, a fact that has been largely forgotten, in part because of Strughold's role in the Luftwaffe's aviation medicine research program during World War II, which included lethal low-pressure experiments on humans. Science and technology studies scholar Jordan Bimm has done fascinating research on this topic; see: Jordan Bimm, "What's in the Mars Jar? Cold War Astrobiology and the Idea of Mars as a Microbial Place," American Anthropological Association (Denver, Colo.: Nov. 2015).

61 "VARIETIES OF TERRESTRIAL MICROBES" Sagan, *Cosmos*, p. 119.

61 RESEMBLE A FOX TERRIER Henry S. F. Cooper, Jr., *The Search for Life on Mars: Evolution of an Idea* (New York: Holt, Rinehart and Winston, 1980), p. 126.

61 *SCIENCE* ARTICLE IN 1969 N. H. Horowitz, et al., "Sterile Soil from Antarctica: Organic Analysis," *Science*, 164, no. 3,883 (1969), pp. 1,054–1,056.

61 ALARMED HOROWITZ Cooper, *The Search for Life on Mars: Evolution of an Idea*, pp. 100–101.

61 DEEPLY TIED TO THE EXOBIOLOGY ENTERPRISE In addition to his role in WESTEX, Horowitz had been supporting the development of Gulliver. NASA

had asked Levin to pick a co-experimenter from a list of PhD scientists to get continued support for the instrument's development, as Levin did not have a doctorate and was seen as more of an engineer than a scientist. He picked Norm Horowitz (and soon began working in his spare time on a doctorate at Johns Hopkins University, which he would earn in just three years). Jay Gallentine, *Infinity Beckoned: Adventuring through the Solar System, 1969–1989,* pp. 18–19.

61 AND THE SAMPLE BAKED His idea focused not on carbon respired by the microbes in the form of carbon dioxide but rather on carbon that had been "fixed," assimilated into the cells themselves. The heat would be enough to destroy the cells, turning any carbon in the soil microbes back into a gas, and a positive Geiger counter reading would then serve as a powerful indication of living organisms. This was the Pyrolytic Release (PR) experiment.

61–62 A MODIFIED VERSION OF GULLIVER This was *Viking*'s Labeled Release (LR) experiment.

62 NICKNAMED THE "CHICKEN SOUP" This was *Viking*'s Gas Exchange (GEX) experiment.

62 THE MORE FOOD THE MARTIAN MICROBES GOT Cooper, *The Search for Life on Mars: Evolution of an Idea,* p. 99.

62 POINTLESS TO STERILIZE Norman H. Horowitz, *To Utopia and Back: The Search for Life in the Solar System* (New York: W. H. Freeman and Company, 1986), p. 120.

62 THE VENUSIAN ATMOSPHERE "55 Years Ago: Mariner 2 First to Venus," NASA (Dec. 14, 2017).

62 JUST AS *VIKING* WAS BEGINNING TO BE PLANNED *Viking* grew out of NASA's *Voyager Mars* program, which was planned between 1966 and 1968 but cancelled due to risk and cost. The mission name was then used for NASA's unrelated *Voyager 1* and *Voyager 2* missions to study the outer planets.

62 CROSSTOWN-RIVAL MIT Poundstone, *Carl Sagan: A Life in the Cosmos,* p. 107.

62 GEOLOGY AND GEOPHYSICS DEPARTMENT In 1952, MIT's Department of Geology was renamed the Department of Geology and Geophysics. In 1967, it became the Department of Earth and Planetary Sciences, and in 1983, it became the Department of Earth, Atmospheric, and Planetary Sciences. "MIT History, School of Science," Institute Archives, MIT Libraries, July 2007.

63 "SPECULATIONS CANNOT BE CONFIRMED" Poundstone, *Carl Sagan: A Life in the Cosmos,* p. 107.

63 WRITTEN A LETTER Ibid., pp. 171–173.

63 FORGIVENESS AND FRIENDSHIP Harold Urey, quoted in Bill Sternberg, "The Sagan Files," *Cornell Alumni Magazine* (March/April 2014).

64 REDID THE ENTIRE THING Poundstone, *Carl Sagan: A Life in the Cosmos,* p. 25.

64 SETTLING OF DUST Ibid., pp. 34–35. Vishniac, for his part, had his own creative hypothesis: Perhaps bright dust was accumulating on the surfaces of Martian plants, then gradually being shaken to the ground, revealing the dark leaves beneath. Extravagant, maybe, but rooted in evidence and logic. Both Vishniac and Sagan had an acute appreciation for what was known and what was unknown

and a deep commitment to the idea that evidence could not be ignored—but in the absence of evidence, the possibilities were limitless.

64 "TURNED TOPSY-TURVY" Vishniac, "Letter from Wolf Vishniac to Clinton P. Anderson, United States Senate."

64 "GREATEST EXPERIMENT" Gilbert V. Levin, "The Curiousness of Curiosity," *Astrobiology,* 15, no. 2 (2015), pp. 101–103.

64 THERE WERE FORTY THOUSAND PARTS "Viking Lander: Creating the Science Teams," in Ezell and Ezell, *On Mars: Exploration of the Red Planet, 1958–1978.*

64 SHOEHORNED INTO A SIXTEEN-KILOGRAM "How Viking Instrument Studies Soil Samples for Signs of Life" and "The Viking Biology Project (Art)" in *Mars Viking* (Redondo Beach, Calif.: TRW, 1976), pp. 11–12.

65 "HAVE HIS SAY" Cooper, *The Search for Life on Mars: Evolution of an Idea,* p. 98.

65 $59 MILLION "Viking Lander: Creating the Science Teams," in Ezell and Ezell, *On Mars: Exploration of the Red Planet, 1958–1978.*

65 CLOUDINESS MIGHT RESULT Joshua Lederberg, "Letter to Dr. Richard S. Young, March 15, 1972," Joshua Lederberg Papers, 1904–2008. Located in: Archives and Modern Manuscripts Collection, History of Medicine Division, National Library of Medicine, Bethesda, Md.; MS C 552.

65 REQUIRED CONDITIONS OF GROWTH "Viking Lander: Creating the Science Teams," in Ezell and Ezell, *On Mars: Exploration of the Red Planet, 1958–1978.*

65 DROWN ANY MARTIAN LIFE Cooper, *The Search for Life on Mars: Evolution of an Idea,* pp. 99–100.

65 "THAT I RECAPTURE SOME" Wolf V. Vishniac, quoted in "Viking Lander: Creating the Science Teams," in Ezell and Ezell, *On Mars: Exploration of the Red Planet, 1958–1978.*

65 HE HAD A WEAK ARM Ephraim Vishniac, phone interview by Sarah Johnson (Sept. 8, 2017).

65 HE HAD A STUTTER Ibid.

66 "KNOWING SOMETHING ABOUT THE CHEMISTRY" Ibid.

66 "ALWAYS IN GOOD WEATHER" Zeddie Bowen, quoted in Sheehan and O'Meara, *Mars: The Lure of the Red Planet,* p. 289.

66 "SOME FASCINATING NEW DISCOVERY" Zeddie Bowen, quoted in Ricki Lewis, "Researchers' Deaths Inspire Actions to Improve Safety," *The Scientist* (Oct. 27, 1997).

66 TRANSFERRED HOME TO ROCHESTER Associated Press, "Wolf V. Vishniac, Micro Biologist," *The New York Times* (Dec. 12, 1973).

66 THE FUNERAL Sagan penned Vishniac's obituary. In it, he underscored what "a remarkably honest, kind, tactful, and thoughtful human being" Vishniac had been, how he had "a deep sense of the fundamental problems of biology and, more than almost all of the other scientists of his time, realized the revolutionary significance for science of the search for extraterrestrial life." Carl Sagan, "Wolf Vladimir Vishniac: An Obituary," *Icarus,* vol. 22, issue 3 (1974), pp. 397–398.

67 READY FOR DELIVERY "Viking Lander: Creating the Science Teams," in Ezell and Ezell, *On Mars: Exploration of the Red Planet, 1958–1978.*

67 GAS FOR FORTY HOURS Eric Burgess, *To the Red Planet* (New York: Columbia University Press, 1978), p. 63.

67 EVEN TRACES OF TRANSIENT LIQUID WATER We now have a great deal of evidence for transient liquid brines on the surface on Mars; see: F. Javier Martín-Torres, María-Paz Zorzano, Patricia Valentín-Serrano, Ari-Matti Harri, Maria Genzer, Osku Kemppinen, Edgard G. Rivera-Valentin, et al., "Transient Liquid Water and Water Activity at Gale Crater on Mars," *Nature Geoscience,* 8, no. 5 (2015), p. 357.

67 FROM PASSING CLOUDS "The Viking Landing Sites: The Questions They'll Answer," in *Mars Viking* (Redondo Beach, Calif.: TRW, 1976), pp. 14.

68 JUNE OF 1976 "Yuty Crater in Chryse Planitia, Mars," NASA (June 22, 1976).

68 DAMPENING THE CONTRAST "Site Certification—and Landing," in Ezell and Ezell, *On Mars: Exploration of the Red Planet, 1958–1978.*

68 "DON'T UNDERSTAND MARS" "Science: Another Delay for Viking," *Time* (July 19, 1976).

68 TIME WAS SHORT *Viking 1* entered into orbit on June 19, 1976, but soon it was clear a July 4 landing was out of the question. *Viking 2* was scheduled to enter orbit on August 7. At an anniversary event, Tom Young, *Viking*'s mission director, recalled that if *Viking 1*'s "landing was delayed too long there would be a traffic jam at Mars that was beyond our capability to manage." Sam McDonald, "'Viking at 40' Events Revisit a Giant Step in NASA's Journey to Mars," NASA (July 26, 2016).

68 NEXT TWO WEEKS Sagan, *Cosmos,* p. 98; see also "Site Certification—And Landing" in Ezell and Ezell, *On Mars: Exploration of the Red Planet, 1958–1978.*

68 SIXTEEN METERS WIDE "Viking 1 Lander," NASA Space Science Data Coordinated Archive, NSSDCA/COSPAR ID: 1975-075C, NASA.

68 "WAVING HELPLESSLY" Sagan, *Cosmos,* p. 96.

68 THE DESCENT CAPSULE BARRELED "Viking 1 Lander," NASA Space Science Data Coordinated Archive, NSSDCA/COSPAR ID: 1975-075C, NASA.

69 RETROROCKETS The fuel was hydrazine, N_2H_4, which wouldn't form organics as a product of combustion; this would minimize the risk that the landing itself would contaminate the lander's surface science findings.

69 COULD LAND ON ANOTHER PLANET The Soviet *Mars 3* mission landed on Mars and survived for about ninety seconds. The Soviet *Venera* missions also survived an hour or two on the surface on Venus.

69 VELLUM AND NOTEPADS At the "Viking at 40" Symposium in 2016, JPL's Robert Manning remarked, "Viking is an amazing accomplishment. Those of us who do it now have a really hard time conceiving how you could build and design this thing before computers were really popular. You did it by hand, with paper, vellum, notepads." McDonald, "'Viking at 40' Events Revisit a Giant Step in NASA's Journey to Mars," NASA.

69 SMALLER THAN THE ROSE BOWL Rebecca Wright, "Interview with A. Thomas Young," NASA Headquarters Oral History Project, NASA Johnson Space Center History Portal (2013).

69 A HUNDRED METERS ACROSS "Viking Encounter Press Kit," NASA (June 4, 1976), p. 18.

69 CHEERILY SINGING This story about Gerry Soffen, Viking's project scientist, was recounted by Gus Guastaferro in "40 Years Remembered: a Shared Experience in Viking Project Leadership," Viking at 40 Symposium Lectures (2016). He may have been referring to the Irving Berlin song with the lyrics, "Blue skies smiling at me; Nothing but blue skies do I see."

69 COLOR ON THE LANDER'S FACSIMILE CAMERAS Elliott C. Levinthal, William Green, Kenneth L. Jones, and Robert Tucker, "Processing the Viking Lander Camera Data," *Journal of Geophysical Research* 82, no. 28 (1977), pp. 4,412–4,420.

69 DISCOVERED THAT THE SKY Poundstone, *Carl Sagan: A Life in the Cosmos,* p. 207.

69 EAGERLY STUDIED THEM Ibid., pp. 204–205.

70 "CHOSEN DULL PLACES" Sagan, *Cosmos,* p. 98.

70 "BIG JOE" "Big Joe in the Chryse Planitia," NASA JPL (Feb. 27, 1997).

70 ROCKS HAD GOTTEN UP Of course, no Martian macrobes ever ambled by, despite special computer techniques designed to reveal movement or changes in scene and search for objects that emitted light in the night. A year later, a *Journal of Geophysical Research* paper concluded that "no evidence, direct or indirect, has been obtained for macroscopic biology on Mars." David McNab and James Younger, *The Planets* (New Haven: Yale University Press, 1999), p. 193; Elliott C. Levinthal, Kenneth L. Jones, Paul Fox, and Carl Sagan, "Lander imaging as a detector of life on Mars," *Journal of Geophysical Research,* 82, no. 28 (1977), pp. 4,468–4,478.

70 ROCK NAMED "SHADOW" McNab and Younger, *The Planets,* p. 191.

70 "HOPPERS ON AN ELECTRIC TRAIN" Sagan, *Cosmos,* p. 102.

70 "WE SENT OUT AND GOT CHAMPAGNE, CIGARS" Gil Levin, "The Viking Labeled Release Biology Experiments," Viking at 40 Symposium Lectures (2016).

70 SOLEMNLY SAT DOWN Cooper, *The Search for Life on Mars: Evolution of an Idea,* p. 129.

70 GAS CHROMATOGRAPH-MASS SPECTROMETER Ibid., pp. 161–169.

71 NEARLY EVERYONE CONCLUDED Ibid., pp. 223–240. One notable exception was Gil Levin, the lead of the Labeled Release experiment. See Levin, "The Viking Labeled Release Experiment and Life on Mars," *Instruments, Methods, and Missions for the Investigation of Extraterrestrial Microorganisms,* vol. 3111, International Society for Optics and Photonics (1997), pp. 146–161.

71 HOROWITZ DECLARED Robert Markley, *Dying Planet: Mars in Science and the Imagination* (Durham, N.C.: Duke University Press, 2005), p. 258.

71 "PLACIDLY MUNCHING" Walter Sullivan, "How to Search for Undefined 'Life' on Mars," *The New York Times* (August 1, 1976).

71 GREAT SAND DUNES NATIONAL MONUMENT Peter Ward, *Life As We Do Not Know It: The NASA Search for (and Synthesis of) Alien Life* (New York: Penguin, 2007), p. 177. Great Sands Dunes National Monument was re-designated as a national park and national preserve in 2000.

72 NOT EVEN SAGAN Cooper, *The Search for Life on Mars: Evolution of an Idea,* p. 122.

Chapter 5: Stone from the Sky

73 WHAT HE KNEW ABOUT BIOLOGY When the Viking payload was being designed, no one realized that the vast majority of microbes wouldn't grow in a nutrient-rich broth—in other words, that they couldn't be cultured. It wasn't until we started identifying life by its genes—not waiting for cells to grow in a petri dish but instead breaking them open within a pinch of dirt or a drop of ocean—that we came to understand that less than 1 percent of the life on Earth could be grown in a lab or, by extension, a lab on a spacecraft on Mars.

73 THE "LIFELESS" SOILS Edward Clinton Ezell and Linda Neuman Ezell, *On Mars: Exploration of the Red Planet, 1958–1978* (Washington, D.C.: The NASA History Series, 1984), p. 236.

73 ALSO COMPLEX EUKARYOTES Helen S. Vishniac and Walter P. Hempfling, "*Cryptococcus Vishniacii* sp. nov., an Antarctic Yeast," *International Journal of Systematic and Evolutionary Microbiology,* 29, no. 2 (1979), pp. 153–158.

73 "SAMPLES FOR IMRE FRIEDMANN" Peter T. Doran, W. Berry Lyons, and Diane M. McKnight, *Life in Antarctic Deserts and Other Cold Dry Environments: Astrobiological Analogs* (Cambridge University Press, 2010), pp. 2–3.

74 PUBLISHED HIS RESULTS E. Imre Friedmann and Roseli Ocampo, "Endolithic Blue-Green Algae in the Dry Valleys: Primary Producers in the Antarctic Desert Ecosystem," *Science,* 193, no. 4,259 (1976), pp. 1,247–1,249.

74 A RICH AND INTRICATE ECOSYSTEM Richard A. Kerr, "Seawater and the Ocean Crust: The Hot and Cold of It," *Science,* 200, no. 4,346 (1978), pp. 1,138–1,187.

74 *THERMUS AQUATICUS* The discovery of a heat-tolerant polymerase within *Thermus aquaticus* cells would lay the foundation for modern molecular biology. The finding paved the way for the development of the polymerase chain reaction, or PCR, which enabled double-stranded DNA to unzip at high temperatures and be replicated, amplifying a tiny signal millions of times. Kary B. Mullis, "The Polymerase Chain Reaction (Nobel Lecture)," *Angewandte Chemie,* 33, 12 (1994), pp. 1,209–1,213; A. Chien, D. B. Edgar, J. M. Trela, "Deoxyribonucleic Acid Polymerase from the Extreme Thermophile *Thermus Aquaticus,*" *Journal of Bacteriology,* 127 (3) (1976), p. 1,550–1,557; Thomas D. Brock and Hudson Freeze, "*Thermus aquaticus* gen. n. and sp. n., a Nonsporulating Extreme Thermophile," *Journal of Bacteriology,* 98, no. 1 (1969).

75 *PSEUDOMONAS BATHYCETES* R. Y. Morita, "Survival of Bacteria in Cold and Moderate Hydrostatic Pressure Environments with Special Reference to Psychrophilic and Barophilic Bacteria" (1976), pp. 279–298; R. G. Gray and J. R. Postgate, eds., "The Survival of Vegetative Microbes" (Cambridge University Press: 1976).

75 DEINOCOCCUS RADIODURANS M.T. Hansen, "Multiplicity of Genome Equivalents in the Radiation-Resistant Bacterium *Micrococcus Radiodurans*," *Journal of Bacteriology*, 134, no. 1 (1978), pp. 71–75; Bevan E. B. Moseley, "Photobiology and Radiobiology of *Micrococcus (Deinococcus) Radiodurans*," in *Photochemical and Photobiological Reviews* (Boston: Springer, 1983), pp. 223–274; Julia M. West and Ian G. McKinley, "The Geomicrobiology of Nuclear Waste Disposal," *MRS Online Proceedings Library Archive*, 26 (1983).

75 SMALL DARK SPOT For a fascinating description of the unfolding of ALH84001 story, from which many of this chapter's details are drawn, see: Kathy Sawyer, *The Rock From Mars: A Detective Story on Two Planets* (New York: Random House, 2006).

75 ABLATES AWAY W. A. Cassidy, E. Olsen, and K. Yanai, "Antarctica: a Deep-Freeze Storehouse for Meteorites," *Science*, 198, no. 4318 (1977), pp. 727–731.

75 ALH84001 TRAVELED BACK Sawyer, *The Rock from Mars: A Detective Story on Two Planets*, p. 20.

75 LIKELY FROM VESTA Ibid., pp. 49–50.

76 NEARBY BUILDING 31 Mimi Swartz, "It Came from Outer Space," *Texas Monthly* (Nov. 1, 1996).

76 SYSTEMATIC STUDY OF FRAGMENTS Sawyer, *The Rock from Mars: A Detective Story on Two Planets*, pp. 51, 58–59.

76 "WHY CHOOSE THIS" John F. Kennedy, "Address at Rice University on the Nation's Space Effort," John. F. Kennedy Presidential Library and Museum (Sept. 12, 1962).

77 "FAIRY TALES EVEN PROBABLE" W. F. Foshag, "Problems in the Study of Meteorites," *The American Mineralogist*, 26, no. 3 (1941), p. 137.

77 THE NAME "THUNDERSTONES" Megan Garber, "Thunderstone: What People Thought About Meteorites Before Modern Astronomy," *The Atlantic* (Feb. 15, 2013).

77 "WORLD FRAGMENTS" Foshag, "Problems in the Study of Meteorites," *The American Mineralogist*, p. 137.

77 METEORITE CALLED SNC Sawyer, *The Rock from Mars: A Detective Story on Two Planets*, p. 53.

77 SHERGOTTY The town in India is now known as Sherghati.

77 ON A DOG T. E Bunch and Arch M. Reid. "The Nakhlites Part I: Petrography and Mineral Chemistry." *Meteoritics* 10, no. 4 (1975), pp. 303–315.

77 THE DIRECT MEASUREMENTS D. D. Bogard and P. Johnson, "Martian Gases in an Antarctic Meteorite?" *Science*, 221 (1983), pp. 651–654; see also: Allan H. Treiman, James D. Gleason, Donald D. Bogard, "The SNC Meteorites Are from Mars," *Planetary and Space Science*, 48 (2000), pp. 1,213–1,230.

78 OLDEST ROCK The date for ALH84001 has since been revised, from 4.51 to 4.091 billion years, and a Martian meteorite that was recently discovered in the Sahara, nicknamed Black Beauty, has replaced ALH84001 as the oldest Martian meteorite (it dates to 4.4 billion years). T. J. Lapen, M. Righter, A. D. Brandon, Vinciane Debaille, B. L. Beard, J. T. Shafer, and A. H. Peslier, "A Younger

Age for ALH84001 and Its Geochemical Link to Shergottite Sources in Mars," *Science,* 328, no. 5976 (2010), pp. 347–351; M. Humayun, Alexander Nemchin, B. Zanda, R. H. Hewins, Marion Grange, Allen Kennedy, J. P. Lorand, et al., "Origin and Age of the Earliest Martian Crust from Meteorite NWA 7533," *Nature,* 503, no. 7477 (2013), p. 513.

78 SIXTEEN MILLION YEARS AGO D. D. Bogard, "Exposure-Age-Initiating Events for Martian Meteorites: Three or Four?" Lunar and Planetary Science Conference, 26 (1995).

78 LISTEN TO ENYA Sawyer, *The Rock from Mars: A Detective Story on Two Planets,* p. 103.

78 ORANGEY KNOBS OF CARBONATE David S. McKay, Everett K. Gibson, Kathie L. Thomas-Keprta, Hojatollah Vali, Christopher S. Romanek, Simon J. Clemett, Xavier D. F. Chillier, Claude R. Maechling, and Richard N. Zare, "Search for Past Life on Mars: Possible Relic Biogenic Activity in Martian Meteorite ALH84001," *Science,* 273, no. 5277 (1996), pp. 924–930.

78 TINY COMPASSES J. L. Kirschvink, "South-Seeking Magnetic Bacteria," *Journal of Experimental Biology,* 86, no. 1 (1980), pp. 345–347; Wei Lin, Dennis A. Bazylinski, Tian Xiao, Long-Fei Wu, and Yongxin Pan, "Life with Compass: Diversity and Biogeography of Magnetotactic Bacteria," *Environmental Microbiology,* 16, no. 9 (2014), pp. 2,646–2,658.

78 ALSO EXTRAORDINARILY PURE Kathie L. Thomas-Keprta, et al., "Magnetofossils from Ancient Mars: A Robust Biosignature in the Martian Meteorite ALH84001," *Applied and Environmental Microbiology,* 68, no. 8 (2002), pp. 3,663–3,672.

80 TRYING TO BE CASUAL Swartz, "It Came from Outer Space," *Texas Monthly.*

80 INCLUDING CARL SAGAN Sawyer, *The Rock from Mars: A Detective Story on Two Planets,* p. 111.

82 TUCKED A SKYPAGER Ibid., p. 137.

83 THE STORY OF HIS ROCK WAS EVERYWHERE Ibid.

83 PRESIDENT CLINTON STEPPED William Jefferson Clinton, "Statement Regarding Mars Meteorite Discovery," Office of the Press Secretary (Aug. 7, 1996).

83 THE FRONT ROW Sawyer, *The Rock from Mars: A Detective Story on Two Planets,* p. 186.

83 "UNBELIEVABLE DAY" Keay Davidson, "Romancing the Red Planet," *San Francisco Examiner* (Aug. 8, 1996).

83 TWO HUNDRED TO TWO THOUSAND DOLLARS Swartz, "It Came from Outer Space," *Texas Monthly.*

84 EXTREMELY CAREFUL SCIENTIST John Noble Wilford, "Clues in Meteorite Seem to Show Signs of Life on Mars Long Ago," *The New York Times* (Aug. 7, 1996).

84 BIG ENOUGH TO ENCAPSULATE Space Studies Board and National Research Council, *Size Limits of Very Small Microorganisms: Proceedings of a Workshop* (Washington, D.C.: National Academies Press, 1999).

84 OWN BROTHER GORDON Matt Crenson, "After 10 Years, Few Believe Life on Mars," *The Washington Post* (Aug. 5, 2006).

84 "FURTHER STRENGTHENED" David S. McKay, Kathy L. Thomas-Keprta, Simon J. Clemett, Everett K. Gibson Jr, Lauren Spencer, and Susan J. Wentworth, "Life on Mars: New Evidence from Martian Meteorites," *Instruments and Methods for Astrobiology and Planetary Missions XII*, vol. 7441 (International Society for Optics and Photonics, 2009), p. 744,102. See also: Kathie L. Thomas-Keprta, et al., "Origins of Magnetite Nanocrystals in Martian Meteorite ALH84001," *Geochimica et Cosmochimica Acta*, 73 (2009), pp. 6,631–6,677; Everett K. Gibson Jr., David S. McKay, Kathie L. Thomas-Keprta, S. J. Wentworth, F. Westall, Andrew Steele, Christopher S. Romanek, M. S. Bell, and J. Toporski, "Life on Mars: Evaluation of the Evidence within Martian Meteorites ALH84001, Nakhla, and Shergotty," *Precambrian Research*, 106, no. 1–2 (2001), pp. 15–34.

85 "A LITTLE TESTY" Crenson, "After 10 Years, Few Believe Life on Mars," *The Washington Post*.

85 KIMONOS FROM THE TIME Sawyer, *The Rock from Mars: A Detective Story on Two Planets*, p. 42; William K. Stevens, "A 'Mellow' Scientist David Stewart McKay," *The New York Times* (August 9, 1996).

85 QUADRUPLE BYPASS HEART SURGERY Sawyer, *The Rock from Mars: A Detective Story on Two Planets*, p. 236; Marc Kaufman, *First Contact: Scientific Breakthroughs in the Hunt for Life Beyond Earth* (New York: Simon and Schuster, 2011), p. 107.

85 "EXTRAORDINARY CLAIMS REQUIRE" Carl Sagan, quoted in Patrizio E. Tressoldi, "Extraordinary Claims Require Extraordinary Evidence: The Case of Non-Local Perception, a Classical and Bayesian Review of Evidences," *Frontiers in Psychology*, 2 (2011), p. 117.

85 "YET EXTRAORDINARY ENOUGH" Carl Sagan, *Billions and Billions: Thoughts on Life and Death at the Brink of the Millennium* (New York: Ballantine, 1998), p. 60.

85 BIOLOGY'S FIVE KINGDOMS . . . DISSOLVED Of course this had happened before—for centuries, living things had been classified simply as plants or animals.

85–86 REPLACED BY A SYSTEM Carl R. Woese, Otto Kandler, and Mark L. Wheelis, "Towards a Natural System of Organisms: Proposal for the Domains Archaea, Bacteria, and Eucarya," *Proceedings of the National Academy of Sciences*, 87, no. 12 (1990), pp. 4,576–4,579.

86 SIMPLE, SINGLE-CELLED ORGANISMS These microbes dominated life on Earth for most of our history. Multicellular organisms are thought to have evolved from unicellular ancestors at least twenty-five times, with a major burst of diversification 600–700 million years ago. Richard K. Grosberg and Richard R. Strathmann, "The Evolution of Multicellularity: A Minor Major Transition?" *Annual Review of Ecology, Evolution, and Systematics*, 38 (2007), pp. 621–654.

86 DOLLY THE SHEEP Keith H. S. Campbell, Jim McWhir, William A. Ritchie, and Ian Wilmut, "Sheep Cloned by Nuclear Transfer from a Cultured Cell Line," *Nature*, 380, no. 6569 (1996), p. 64.

86 COUNTLESS TONS OF ROCKS During the Late Heavy Bombardment, planetary bodies in our solar system were impacted with a higher amount of material; see: William F. Bottke and Marc D. Norman, "The Late Heavy Bombardment," *Annual Review of Earth and Planetary Sciences*, 45 (2017), pp. 619–647.

87 WHAT DAVID MCKAY HOPED Kaufman, *First Contact: Scientific Breakthroughs in the Hunt for Life Beyond Earth*, p. 107.

Chapter 6: Traversing

88 SIZE OF A SUITCASE "Press Kit: Mars Pathfinder Landing," NASA (July 7, 1997).

89 "FASTER, BETTER, CHEAPER" Howard E. McCurdy, *Faster, Better, Cheaper: Low-Cost Innovation in the U.S. Space Program* (Baltimore: JHU Press, 2001).

89 THE SOUTH BRONX William J. Broad, "Scientist at Work: Daniel S. Goldin, Bold Remodeler of a Drifting Agency," *The New York Times* (Dec. 21, 1993).

89 A FIFTEENTH THE COST McCurdy, *Faster, Better, Cheaper*.

89 BARREL STRAIGHT INTO A PLANET John Noble Wilford, "More Than 20 Years After Viking, Craft Is to Land, and Bounce, on Mars," *The New York Times* (July 1, 1997).

89 TWENTY-METER KEVLAR TETHER David R. Williams, "Mars Pathfinder Atmospheric Entry Strategy," NASA Goddard Space Flight Center (Dec. 30, 2004).

89 SURFACE OF A NEW WORLD "Mars Pathfinder Transmits Dramatic Color Images," CNN (July 5, 1997).

90 FEW HUNDRED METERS Mars Pathfinder Frequently Asked Questions, NASA (April 10, 1997).

90 WHERE ABOUT 80 PERCENT "Rover 'Holds Hands' with Barnacle Bill," CNN (July 7, 1997).

90 ON ROCKER-BOGIES "Mars Curiosity Rover: Wheels and Legs," NASA.

90 FROM FAR-OFF PLACES D. M. Nelson and R. Greeley, "Xanthe Terra Outflow Channel Geology at the Mars Pathfinder Landing Site," *Journal of Geophysical Research Planets*, 104, no. 4 (1999), pp. 8,653–8,669.

90 "BARNACLE BILL" R. Rieder, T. Economou, H. Wänke, A. Turkevich, J. Crisp, J. Brückner, G. Dreibus, and H. Y. McSween, "The Chemical Composition of Martian Soil and Rocks Returned by the Mobile Alpha Proton X-Ray Spectrometer: Preliminary Results from the X-Ray Mode," *Science*, 278, no. 5344 (1997), pp. 1,771–1,774.

90 SOLIDIFYING, AND REMELTING John Noble Wilford, "Mars History: Heat and Cold Leave Marks," *The New York Times* (July 9, 1997).

91 AFTER CARTOON CHARACTERS While these cartoon character names didn't raise eyebrows in the late 1990s, a decade later, when a fairy-tale naming scheme was chosen during the Mars *Phoenix* mission, NASA's legal office realized that NASA could be sued for using copyrighted material. Only names in the public domain were thenceforth allowed. Rod Pyle, *Destination Mars: New Explorations of the Red Planet* (Amherst, N.Y.: Prometheus Books, 2012), p. 238.

91 CRESTING INTO DUNES Ronald Greeley, Michael Kraft, Robert Sullivan, Gregory Wilson, Nathan Bridges, Ken Herkenhoff, Ruslan O. Kuzmin, Michael Malin, and Wes Ward, "Aeolian Features and Processes at the Mars Pathfinder

Landing Site," *Journal of Geophysical Research: Planets,* 104, no. E4 (1999), pp. 8,573–8,584.

92 "BACK ON MARS" Matt Crenson, "Back on Mars," *The Courier-Journal* (Louisville, Ky., July 5, 1997).

92 CLOSED WITH A QUOTE John Noble Wilford, "Scientists Await Craft's Plunge to Mars Today," *The Courier-Journal* (Louisville, Ky., July 4, 1997).

93 "MOST IMPORTANT NUMBER" Mars Pathfinder Science Results: Rotational and Orbital Dynamics, NASA; see also: W. M. Folkner, et al., "Interior Structure and Seasonal Mass Redistribution of Mars from Radio Tracking of Mars Pathfinder," *Science,* 278, no. 5,344 (1997), pp. 1,749–1,752.

94 THICKER ATMOSPHERE A magnetic field also deflects charged particles in the solar wind, reducing atmospheric loss to space. Even though Venus lacks a dynamo—it's rotating too slowly—gravity has helped it to hold on to its dense atmosphere. Mars, with its weaker gravity, never had as tight a grip.

94 "LAST GREAT CHALLENGE" Bertrand Piccard and Brian Jones, *Around the World in 20 Days: The Story of Our History-Making Balloon Flight* (Hoboken, N.J.: John Wiley & Sons, 1999).

94 CHINA, IRAQ, AND LIBYA Malcolm W. Browne, "Balloon Soars Over Atlantic, Setting Record in Solo Flight," *The New York Times* (Aug. 12, 1998).

94 FOSSETT WAS Steve Fossett biography, National Aviation Hall of Fame.

95 ANNOUNCED TO *THE NEW YORK TIMES* "Signals from Mars from a Balloon," *The New York Times* (May 2, 1909).

95 TESLA, CAPTIVATED Out on the plains of the Colorado Plateau, Tesla had constructed a laboratory in a barn-like building with a twenty-four-meter tower, with a sign on the door saying, GREAT DANGER, KEEP OUT. He spent night after night there watching storms in the darkness, detecting lightning strikes with his new equipment. Once he even began picking up what he assumed to be faint extraplanetary signals—disturbances he concluded were coming from Mars—which happened to disappear as Mars set over the horizon. It may have been a Jovian storm, or perhaps even his rival Marconi, who was simultaneously attempting to send pulses across the Atlantic from Cornwall to Canada using an early spark gap transmitter. See: W. Bernard Carlson, *Tesla: Inventor of the Electrical Age* (Princeton, N.J.; Princeton University Press, 2013), p. 277; Marc Seifer, *Wizard: The Life and Times of Nikola Tesla: Biography of a Genius* (New York: Citadel Press, 1998), p. 217.

95 ONLY FIVE MINUTES AWAY "That Prospective Communication with Another Planet," *Current Opinion,* March 1919, p. 170; Michael Brown, "Radio Mars: The Transformation of Marconi's Popular Image, 1919–1922," in *Transmitting the Past: Historical and Cultural Perspectives on Broadcasting,* ed. J. Emmett Winn and Susan Lorene Brinson (University of Alabama Press, 2005), p. 23.

95 "ALUMINUM FOR LIGHTNESS" "Signals from Mars from a Balloon," *The New York Times.*

95 THE *MASSACHUSETTS* "Offer Balloon to Todd," *The New York Times* (May 7, 1909).

95 "THE BIG LISTEN" Todd had lobbied for no less than a worldwide radio silence to try to detect signals from Mars. This, he believed, would be the only way to address the fact that radio traffic had begun to swamp the airwaves, complicating his listening attempts. During "the Big Listen," Todd even enlisted the help of Francis Jenkins, the man who would go on to open the first television broadcasting station in America. Jenkins had invented a radio telescope of sorts, which he called a "radio photo message continuous transmission machine." The apparatus's nine-meter-long swath of chemical-treated film was designed for the picturization of radio signals and would be the perfect way to create the first human record of Martian communication. That August, the film Jenkins's machine produced at 1515 Connecticut Avenue showed, in "black on white, everything that was 'picked up' out of the air in about twenty-nine hours with a receiving apparatus adjusted to a wave length of six thousand meters." Three and a half meters of footage—a strange series of dots and dashes—was turned over to the chief of the code section in the office of the chief signal officer of the Army, and the rest to the chief of the radio division of the Bureau of Standards. Todd, surely recognizing that he had failed to demonstrate the existence of an alien civilization capable of communicating with Earth by radio, nevertheless reserved judgment on the results and hailed the experiment as a breakthrough for the scientific process. "We now have a permanent record which can be studied," he said, "and who knows, until we have studied it, just what these signals may have been? But the important thing is to have a record." He continued: "Three years ago Marconi was reported as saying he had heard signals from Mars. A few days ago he was quoted as saying he was too busy to listen to possible messages from Mars and that it was a ridiculous idea to do so. He changed his mind, and no one knows what he heard the first time. With our photograph, however, it is not a question of what one man heard. It is a permanent record, which all can study." He went on, "The Jenkins machine is perhaps the hypothetical Martians' best chance of making themselves known to Earth. If they have, as well they may, a machine that now is transmitting earthward their 'close-up' of faces, scenes, buildings, landscapes, and whatnot, their sunlight values having been converted into electric values before projection earthward, all these would surely register on the weirdly unique little mechanism." See: "Weird 'Radio Signal' Film Deepens Mystery of Mars," *The Washington Post* (Aug. 27, 1924); Craig P. Bauer, *Unsolved!: The History and Mystery of the World's Greatest Ciphers from Ancient Egypt to Online Secret Societies* (Princeton, N.J.: Princeton University Press, 2017), pp. 500–503.

95 AUDOUIN DOLLFUS He worked at the Meudon Observatory in Paris and nearly convinced Carl Sagan to accept a postdoctoral fellowship with him in lieu of going to Berkeley. Sheehan and O'Meara, *Mars: The Lure of the Red Planet*, p. 233; Ray Spangenburg, Kit Moser, and Diane Moser, *Carl Sagan: A Biography* (Westport, Conn.: Greenwood Publishing Company, 2004), p. 42.

96 WEATHER BALLOONS, CLUSTERED Pierre de Latil and Tom Margerison, "Planetary Observations by the Multi-Balloon Technique," *The New Scientist* (May 7, 1959); Louis de Gouyon Matignon, "Audouin Dollfus, The French Aeronaut," *Space Legal Issues* (May 25, 2019).

96 SLUNG BENEATH De Latil and Margerison, "Planetary Observations by the Multi-Balloon Technique," *The New Scientist*.

96 STRANGE HAZARD Ibid.

96 STRING OF SPANISH ONIONS Ibid.

96 WROTE IN HIS LOGBOOK Mark Karpel, "The Drifters," *Air & Space* (Aug. 2010).

96 FOURTEEN THOUSAND METERS Matignon, "Audouin Dollfus, The French Aeronaut," *Space Legal Issues;* Sheehan and O'Meara, *Mars: The Lure of the Red Planet,* p. 233.

96 FAILED TO GET HIS MEASUREMENT Dollfus was able to make other measurements of Venus and the moon on that flight. A couple of years later, at a spot high in the Swiss Alps, Dollfus tried again for Mars, setting up a special spectroscope that allowed him to separate the signal from Mars by its Doppler shift, and with it he finally was able to secure a positive detection. He calculated that if all the water in the Martian atmosphere condensed to the surface, it would form a layer less than eight one-thousandths of an inch thick. As happy as he was that he'd finally succeeded in making the observation, he also knew what it meant: that Mars was many times drier than the most parched places on Earth, and that it would be challenging, perhaps very challenging, for life-forms to inhabit its surface. Sheehan and O'Meara, *Mars: The Lure of the Red Planet,* p. 233; Matignon, "Audouin Dollfus, The French Aeronaut," *Space Legal Issues.*

97 OUR AEROBOT PAYLOAD Raymond E. Arvidson, et al., "Aerobot Measurements Successfully Obtained During Solo Spirit Balloon Mission," *EOS,* 80, no. 14 (1999), pp. 153, 158–159.

97 AEROBOT HAD SENSORS Tony Fitzpatrick, "NASA Payload Part of Cargo on Solo Spirit," *Record,* 22, no. 15 (Dec. 11, 1997).

97 STADIUM IN MENDOZA Tribune News Services, "Fossett Lifts Off on 4th Balloon Attempt to Circle the World," *Chicago Tribune* (Aug. 8, 1998).

97 THE LATEST TECHNOLOGIES Steve Mills, "Balloonist Charting a High-Tech Course," *Chicago Tribune* (Jan. 19, 1997).

97 SPECIAL ROZIÈRE DESIGN Michelle Knott, "Technology: Up, up and Around the World," *New Scientist* (Dec. 21, 1996).

98 EVERY TEN SECONDS Arvidson, et al., "Aerobot Measurements Successfully Obtained During Solo Spirit Balloon Mission," *EOS,* pp. 153–159.

98 WAS LEFT BOBBING Malcolm W. Browne, "Balloonist to Take It Easy After Stormy Crash at Sea," *The New York Times* (Aug. 18, 1998).

98 HE ACTIVATED THAT BEACON TWICE Jon Jeter, "Storm Ends Balloonist's Quest in Coral Sea," *The Washington Post* (Aug. 17, 1998).

99 AUSTRALIAN YACHTSMAN Rohan Sullivan, "Balloonist Fossett Rescued from Sea," AP News (August 17, 1998); "Balloonist Rescued Off Australia," CNN (August 17, 1998).

99 RELYING ON OXYGEN TANKS AND BARELY SLEEPING Steve Fossett interview, Public Broadcasting Corporation, *NOVA.*

99 SINGED HIS EYEBROWS Jeter, "Storm Ends Balloonist's Quest in Coral Sea," *The Washington Post.*

99 "SMELL THE ROSES" Browne, "Balloonist to Take It Easy After Stormy Crash at Sea," *The New York Times.*

Chapter 7: Periapsis

102 KALEIDOSCOPIC NEW MAP The new map was presented at the American Geophysical Union Conference in December 1998; see: M. T. Zuber, D. E. Smith, J. B. Garvin, D. O. Muhleman, S. C. Solomon, H. J. Zvally, G. A. Neumann, O. Aharonson, and A. Ivanov, "Geometry of the North Polar Icecap of Mars from the Mars Orbiter Laser Altimeter," American Geophysical Union Conference (1998). It was published the same month; see: Maria T. Zuber, David E. Smith, Sean C. Solomon, James B. Abshire, Robert S. Afzal, Oded Aharonson, Kathryn Fishbaugh, et al., "Observations of the North Polar Region of Mars from the Mars Orbiter Laser Altimeter," *Science,* 282, no. 5,396 (1998), pp. 2,053–2,060. Also see: "New View of Mars' North Pole Reported in *Science,*" *EurekAlert!* (Dec. 6, 1998).

103 SMOOTH AS THE ABYSSAL PLAINS Oded Aharonson, Maria T. Zuber, and Daniel H. Rothman, "Statistics of Mars' Topography from the Mars Orbiter Laser Altimeter: Slopes, Correlations, and Physical Models." *Journal of Geophysical Research: Planets* 106, no. E10 (2001), pp. 23723–23735.

103 "STRANGE, LARGE OCEANIC FISHES" "Background of the MOLA Investigation: Background and General Information," MOLA Science Investigation, NASA Goddard Space Flight Center.

103 "PLANETARY OBSERVERS" "The '80s > Mars Observer," NASA JPL; Michael C. Malin, et al., "An Overview of the 1985–2006 Mars Orbiter Camera Science Investigation," *Mars: The International Journal of Mars Science and Exploration,* 5 (2010), pp. 1–60.

103 AMIDST THE COAL FIELDS MIT News Office, "3Q: Maria Zuber, Daughter of Coal Country," *MIT News* (Feb. 27, 2017).

104 SCENES OF MISSION CONTROL "Maria T. Zuber," YouTube video, posted by MIT Infinite History (April 8, 2016).

104 LIEUTENANT UHURA Maria's GRAIL mission to the moon would launch on the forty-fifth anniversary of *Star Trek,* her favorite television show. Nichelle Nichols, the woman who played Lieutenant Uhura, would be there to celebrate. When she took on the role in 1966, Nichols was one of the first African American actresses portrayed on television in a non-menial role (Martin Luther King once told her *Star Trek* was the only television show he let his children watch, and it was because of her powerful role).

104 SCRIMPED AND SAVED "Maria Zuber: The Geophysicist Became the First Woman to Lead a NASA Planetary Spacecraft Mission," *Physics Today* (June 27, 2017).

104 FOR HOURS Maria Zuber, personal interview by Sarah Johnson (Cambridge, Mass.: May 1, 2019).

104 WITH FIVE CHILDREN Ibid.

104 UNIVERSITY OF PENNSYLVANIA "Maria T. Zuber," YouTube video, MIT Infinite History.

104 FRANKLIN INSTITUTE Chandler, "In Profile: Maria Zuber," *MIT News.*

104 MODELS OF PLANETARY EVOLUTION As a graduate student, Maria worked on models of fluid dynamics, often involving highly nonlinear viscous fluids; e.g., see: M. T. Zuber and E. M. Parmentier, "A Geometric Analysis of Surface Deformation: Implications for the Tectonic Evolution of Ganymede," *Icarus,* 60, no. 1 (1984), pp. 200–210; M. T. Zuber, "A Dynamic Model for Ridge Belts on Venus and Constraints on Lithospheric Structure," Lunar and Planetary Science Conference, 17 (1986).

105 A CHEAPER VERSION "Maria T. Zuber," YouTube video, MIT Infinite History.

105 FINAGLED A SECURITY CLEARANCE Zuber, personal interview by Johnson.

105 IN A REFRIGERATOR LIGHT Bruce Banerdt, "The Martian Chronicles," vol. 1, no. 3, NASA.

105 WHEN NASA EVENTUALLY APPROVED Zuber, personal interview by Johnson.

105 SNAPPING ITS FIRST FAR-OFF PICTURES The MOLA instrument did not itself have a camera, but the spacecraft did. It was a three-component system, with one narrow-angle and two wide-angle cameras. Michael C. Malin, G. E. Danielson, A. P. Ingersoll, H. Masursky, J. Veverka, M. A. Ravine, and T. A. Soulanille, "Mars Observer Camera," *Journal of Geophysical Research: Planets,* 97, no. E5 (1992), pp. 7,699–7,718.

106 RIGHT BEFORE ATMOSPHERIC ENTRY Michael C. Malin, et al., "An Overview of the 1985–2006 Mars Orbiter Camera Science Investigation," *Mars: The International Journal of Mars Science and Exploration.* The communications outage was planned to protect the spacecraft while the propulsion system tanks (which would be used to slow the spacecraft down, allowing it to be captured into orbit) were being pressurized.

106 MARIA WAS HOPEFUL Zuber, personal interview by Johnson.

106 EVERY TWENTY MINUTES John Noble Wilford, "NASA Loses Communication with Mars Observer," *The New York Times* (Aug. 23, 1993).

106 TRYING TO REESTABLISH CONTACT Wilford, "Another Hope to Save Mars Craft Is Dashed," *The New York Times* (Aug. 26, 1993).

106 CLOUDS IN THE PACIFIC Ibid.

106 LIKE A PHANTOM Ben Evans, "And Then Silence: 25 Years Since the Rise and Fall of Mars Observer," *America Space* (Sept.24, 2017).

107 "I HOPE YOU FIND IT" Zuber, personal interview by Johnson.

107 THE PANEL ANNOUNCED Like a great grizzly, the spacecraft had been hibernating, but the communications-satellite technology hadn't been designed to lie dormant for months. In January 1994, the investigation panel from the Naval Research Laboratory concluded that the most likely cause of the spacecraft's disappearance was a ruptured fuel-pressurization tank in its main propulsion system. Hypergolic monomethyl hydrazine may have leaked past valves during the eleven-month journey to Mars and inadvertently mixed with nitrogen tetroxide. The leaking fuel may have induced an extremely high spin rate and likely damaged critical components aboard *Mars Observer* itself. "Mars Observer Mission Failure Investigation Board Report Vol. 1," NASA (Dec. 31, 1993).

107 REFLY THE MISSION Zuber, personal interview by Johnson; Zoe Strassfield, "An Interview with Maria Zuber (Part I)," EAPS (Nov. 14, 2012).

107 THAT KIND OF INSPIRATION Zuber, personal interview by Johnson.

107 LIKE *PATHFINDER Mars Global Surveyor* launched on November 7, 1996, whereas *Pathfinder* launched a month later, on December 4, 1996. Because *Pathfinder* wouldn't need to slow down to enter orbit, it was able to travel on a faster trajectory to Mars. It arrived on July 4, 1997, whereas *Mars Global Surveyor* didn't arrive until September 11, 1997. It was a new tack for NASA: "The lingering *Challenger* disaster and recurring fuel tank problems with the shuttle compounded NASA's negative public image, giving rise to the widespread perception that taxpayer funding for space exploration was being squandered; meanwhile the space science community was growing impatient for continuous new data, and the civilian space budget was dwindling." See: Stephanie A. Roy, "The Origin of the Smaller, Faster, Cheaper Approach in NASA's Solar System Exploration Program," *Journal of Space Policy*, 14 (1998), pp. 153–171.

108 SECOND-IN-COMMAND "Press Kit: Mars Observer," NASA (Sept. 1992).

108 A REPORTER CALLED Zuber, personal interview by Johnson.

108 A TINY LEVER The lever had been part of the damper arm, meant to keep the solar panel from banging shut like a screen door.

108 NO LARGER THAN A CARABINER "Press Kit: Mars Global Surveyor Arrival," NASA (Sept. 1997); Kirk Goodall, "An Explanation of How Aerobraking Works," Mars Global Surveyor, NASA; Diane Ainsworth, "Mars Pathfinder Passes Global Surveyor on Its Way to Mars," Public Information Office, NASA JPL (March 14, 1997).

108 RIGHT INTO THE FIVE-CENTIMETER HINGE Michael C. Malin, et al., "An Overview of the 1985–2006 Mars Orbiter Camera Science Investigation," *Mars: The International Journal of Mars Science and Exploration*, pp. 1–60; Diane Ainsworth, "Mars Global Surveyor to Aerobrake in Modified Configuration," Public Information Office, NASA JPL (Apr. 30, 1997).

108 SHORT OF ITS FULLY OPEN POSITION "Exploring Mars: Mars Global Surveyor Mapped the Red Planet," *SpaceToday* (2007).

108 "AEROBRAKING" Aerobraking had only been tested once. At the end of its mission in 1994, the *Magellan* spacecraft had dipped into the thick haze of Venus's atmosphere as part of an engineering demonstration. The spacecraft burned itself up in the hot clouds, but it had slowed down dramatically. *Mars Global Surveyor* was designed to graze across the top of the much thinner, cooler upper Martian atmosphere. For extra drag, the solar panels had deployable flaps at the tips. Daniel T. Lyons, "Mars Global Surveyor: Aerobraking with a Broken Wing," JPL Technical Report (July 30, 1997); "Press Kit: Mars Global Surveyor Arrival," NASA; Kirk Goodall, "An Explanation of How Aerobraking Works," Mars Global Surveyor, NASA.

108 FACTOR OF FIVE Ibid.

109 WIGGLE THE PANEL "Flight Status Report" (Jan. 24, 1997); Lyons, "Mars Global Surveyor: Aerobraking with a Broken Wing."

109 THE CHAIR TOLD HER Zuber, personal interview by Johnson.

109 "MAY HAVE BROKEN OFF" Ibid.

109 SECRETARY BLANCHED Ibid.

109 PANEL HADN'T IN FACT SNAPPED "Exploring Mars: Mars Global Surveyor Mapped the Red Planet," *SpaceToday.*

110 CORRECT ALIGNMENT Diane Ainsworth, "Surveyor Resumes Aerobraking, Heads for New Mapping Orbit," Public Information Office, NASA JPL (Nov. 10, 1997); Mary Hardin, "Mars Global Surveyor Successfully Completes Aerobraking," Media Relations Office, NASA JPL (Feb. 4, 1999).

110 GIOVANNI SCHIAPARELLI Lovely depictions of Schiaparelli's life and work can be found in: William Sheehan, *The Planet Mars: A History of Observation and Discovery* (Tucson: University of Arizona Press, 1999); William Sheehan, "Giovanni Schiaparelli: Visions of a Colour Blind Astronomer," *Journal of the British Astronomical Association,* 107 (1997), pp. 11–15; Sheehan and O'Meara, *Mars: The Lure of the Red Planet*; pp. 103-123; K. Maria D. Lane, "Mapping the Mars Canal Mania: Cartographic Projection and the Creation of a Popular Icon," *Imago Mundi,* 58:2, (2006), pp. 198–211; and Michele T. Mazzucato, "Giovanni Virginio Schiaparelli," *Journal of the Royal Astronomical Society of Canada,* 100, no. 3 (2006), pp. 114–117.

110 FURNACE OPERATOR Ibid.

110 THE AGE OF FIFTEEN A. Manara and G. Trinchieri, "Schiaparelli and His Legacy," *Memorie della Societa Astronomica Italiana,* 82 (2011), p. 209.

110 AT NINETEEN A. Ferrari, "Between Two Halley's Comet Visits," *Memorie della Societa Astronomica Italiana,* 82, no. 2 (2011), pp. 232–239; Mazzucato, "Giovanni Virginio Schiaparelli," *Journal of the Royal Astronomical Society of Canada.*

111 TEACHING ELEMENTARY MATHEMATICS Manara and Trinchieri, "Schiaparelli and His Legacy," *Memorie della Societa Astronomica Italiana,* pp. 209–218.

111 HE LACKED TOWELS Agnese Mandrino, et al., "Calze, Camicie, Frack e Bottoni Sullo Sfondo del Trattato di Parigi," *Di Pane e Di Stelle* (Aug. 29, 2010); G. V. Schiaparelli, letter dated April 29, 1856, *Historical Archive of the Astronomical Observatory of Brera,* Box 370.

111 GYMNASIUM PORTA NUOVA Manara and Trinchieri, "Schiaparelli and His Legacy," *Memorie della Societa Astronomica Italiana,* p. 209.

111 LATIN, FRENCH, GREEK, AND HEBREW Mandrino, et al., "Natale 1855: Poesia," *Di Pane e Di Stelle* (April 19, 2010); Schiaparelli, diary dated December 25, 1855, *Historical Archive of the Astronomical Observatory of Brera.*

111 "WHAT IS WORSE" Mandrino, et al., "Freddo e Fame Non Lasciano Studiare," *Di Pane e Di Stelle* (April 26, 2010); Schiaparelli, letter dated December 29, 1855, *Historical Archive of the Astronomical Observatory of Brera.*

111 "BARBARIAN COUNTRY" Mandrino, et al., "Prima Della Partenza: Tranquillizzare i Genitori," *Di Pane e Di Stelle* (June 28, 2010); Schiaparelli, letter dated December 18, 1857, *Historical Archive of the Astronomical Observatory of Brera.*

111 PHILOSOPHY, GEOGRAPHY, METEOROLOGY Mazzucato, "Giovanni Virginio Schiaparelli," *Journal of the Royal Astronomical Society of Canada;* Manara and Trinch-

ieri, "Schiaparelli and His Legacy," *Memorie della Societa Astronomica Italiana;* Ferrari, "Between Two Halley's Comet Visits," *Memorie della Societa Astronomica Italiana.*

111 DEVOURED THE CANON P. Tucci, "The Diary of Schiaparelli in Berlin (October 26, 1857–May 10, 1859): A Guide for His Future Scientific Activity," *Memorie della Societa Astronomica Italiana,* 82, no. 2 (2011), pp. 240–247.

111 LEARNING ARABIC One of his younger brothers became a famous professor of Arabic, and that younger brother had a daughter, Else, who became a world-renowned fashion designer.

111 HE CALLED IT HESPERIA Mazzucato, "Giovanni Virginio Schiaparelli," *Journal of the Royal Astronomical Society of Canada,* p. 117.

111 HE CLEAVED THE GLOBE Jürgen Blunck, *Mars and Its Satellites: A Detailed Commentary on the Nomenclature* (Smithtown, N.Y.: Exposition Press, 1982), p. 15; William Sheehan, *The Planet Mars: A History of Observation and Discovery* (Tucson: University of Arizona Press, 1999), p. 73.

112 ARGYRE The legendary silver island at the mouth of Ganges, today's Arakan, Burma.

112 COLUMNS OF HERCULES Or Herculis Columnae, the two promontories that flank the eastern entrance of the Strait of Gibraltar.

112 CHRYSE Or the Land of Gold, described as in the region of Thailand/Malacca by Ptolemy.

112 "BLACK MELANCHOLY" Mandrino, et al., "Problemi di Ieri, Problemi di Oggi," *Di Pane e Di Stelle* (July 26, 2010); G. V. Schiaparelli, letter dated 1910, *Historical Archive of the Astronomical Observatory of Brera;* Schiaparelli had felt loss profoundly, especially the loss of his brother Eugenio, who died from wounds in the Battle of Solferino. He once wrote about how he wished that nations would talk less about Armstrong and Krupp, the makers of guns, and more about Merz, Cooke, and Clark, the makers of telescopes.

112 UPSIDE DOWN "Mars Is Earth, Upside Down," ToponymyMars (June 2, 2013).

112 "A CURIOUS AND DISORDERED ARRANGEMENT" Giovanni Virginio Schiaparelli, *Astronomical and Physical Observations of the Axis of Rotation and the Topography of the Planet Mars: First Memoir, 1877–1878,* trans. William Sheehan (Springfield, Ill.: Association of Lunar and Planetary Observers, 1996), pp. 1, 3.

112 NAMES THAT WOULD FILTER DOWN Schiaparelli's "de-Anglicized nomenclature" provoked deep-seated territorialism among many British astronomers. See the fascinating discussion in: K. Maria D. Lane, "Geographers of Mars: Cartographic Inscription and Exploration Narrative in Late Victorian Representations of the Red Planet," *Isis,* vol. 96, no. 4 (December 2005), pp. 477–506, p. 488 in particular.

112 AS AN ASTRONAUT Zuber, personal interview by Johnson.

113 INSTRUMENTS TO SAFELY DEPLOY The instruments needed to be above the atmosphere, protected from the heat of aerobraking, to safely deploy. "Scientists Get Images of Mars Pole, Clouds," *MIT News* (Dec. 9, 1998); Maria T. Zuber, et al., "Observations of the north polar region of Mars from the Mars Orbiter Laser Altimeter," *Science* 282, no. 5396 (1998), pp. 2,053–2,060. David Spencer

and R.H. Tolson, "Aerobraking Cost/Risk Decisions." *J. Spacecraft and Rockets.* (2007), 44; Greg Mehall, "Mars Global Surveyor and TES Update," *TES News* 7, no. 1 (Jan. 1998).

113 HER POLAR ARTICLE Maria T. Zuber, et al., "Observations of the North Polar Region of Mars from the Mars Orbiter Laser Altimeter," *Science.*

113 A LOW-COST AFFAIR Tony Spear, "NASA FBC Task Final Report," NASA (March 13, 2001).

114 RESULT OF A MIX-UP Arden Albee, Steven Battel, Richard Brace, Garry Burdick, John Casani, Jeffrey Lavell, Charles Leising, Duncan MacPherson, Peter Burr, and Duane Dipprey, "Report on the Loss of the Mars Polar Lander and Deep Space 2 Missions," NASA Technical Report (2000).

114 ULTIMI SCOPULI Eric J. Kolb and Kenneth L. Tanaka, "Accumulation and Erosion of South Polar Layered Deposits in the Promethei Lingula Region, Planum Australe, Mars," *Mars: The International Journal of Mars Science and Exploration,* 2 (2006), pp. 1–9.

114 TOP SPEED The lower gravity helped, but it was still the equivalent of a fall from a four-story building. Kenneth Chang, "Remains of Failed Mars Lander May Have Been Found," *The New York Times* (May 10, 2005).

114 SANDS OF SAMARA "Possible Crash Site of Mars 6 Orbiter/Lander in Samara Vallis," Lunar and Planetary Laboratory, High Resolution Imaging Science Experiment (image acquired May 26, 2007).

114 PENNANT WITH THE STATE EMBLEM The 1971 Soviet *Mars 2* mission likely crashed near 45°S, 313°W; see: "Mars 3 Lander," NASA Space Science Data Coordinated Archive, NSSDCA/COSPAR ID: 1971-049F.

114 BRITISH ASTROBIOLOGY MISSION The *Beagle 2* mission, named after Charles Darwin's famous ship, rode to Mars aboard the European Space Agency's *Mars Express* mission in 2003. *Beagle 2* was designed to search for signs of past life on the Martian surface and in the shallow subsurface, analyzing samples in its miniature onboard laboratory. Its other objectives were characterizing the landing site geology, mineralogy, geochemistry and oxidation state, the physical properties of the atmosphere and surface layers, and collecting data on Martian meteorology and climatology.

114 ONE STILL FOLDED The mission came tantalizingly close to succeeding. Orbital imaging of the site in 2015 using the Mars Reconnaissance Orbiter High Resolution Imaging Science Experiment (HiRISE) appears to show that three of the four solar panels deployed successfully. The fourth panel likely failed or only deployed partially, thereby obscuring the transmitter's antenna. See: J. C. Bridges, et al. "Identification of the Beagle 2 Lander on Mars," *Royal Society Open Science,* 4, no. 10 (2017), pp. 170, 785.

114 CD-ROM, LIKELY SHATTERED Ben Huh, "Kids' Names Going to Mars," *South Florida Sun-Sentinel* (Deerfield Beach, Fla., March 3, 1998); Ashwin R. Vasavada, et al., "Surface Properties of Mars' Polar Layered Deposits and Polar Landing Sites," *Journal of Geophysical Research,* 105, no. E3 (2000), pp. 6,961–6,969.

114 "HOW BAD COULD THIS" Zuber, personal interview by Johnson.

115 MORE DATA THAN ANY PREVIOUS MARS MISSION "Press Kit: Phoenix Landing: Mission to the Martian Polar North," NASA (May 2008).

115 THE TOPOGRAPHICAL MEASUREMENTS THEY RETURNED "Mars Orbiter Laser Altimeter (MOLA) Elevation Map," Goddard Space Flight Center (May 28, 1999).

115 PLATE TECTONICS As mentioned in an earlier endnote, while many researchers agree Mars is likely a "one-plate planet," it has been suggested that Valles Marineris could be a plate boundary. See: D. Breuer and T. Spohn, "Early Plate Tectonics Versus Single-Plate Tectonics on Mars: Evidence from Magnetic Field History and Crust Evolution," *Journal of Geophysical Research: Planets,* 108, no. E7 (2003); An Yin, "Structural Analysis of the Valles Marineris Fault Zone: Possible Evidence for Large-Scale Strike-Slip Faulting on Mars," *Lithosphere,* 4, no. 4 (2012), pp. 286–330.

115 THREE KILOMETER BLANKET OF ROCK "Mars Basher," *Scientific American* (May 31, 1999).

115 WHACKED MARS HARD J. H. Roberts, R. J. Mills, and M. Manga, "Giant Impacts on Early Mars and the Cessation of the Martian Dynamo," *Journal of Geophysical Research Planets,* 114, no. E4 (2009).

115 SMOOTHEST SURFACE Chandler, "In Profile: Maria Zuber," *MIT News;* David E. Smith, Maria T. Zuber, Sean C. Solomon, Roger J. Phillips, James W. Head, James B. Garvin, W. Bruce Banerdt, et al., "The Global Topography of Mars and Implications for Surface Evolution," *Science,* 284, no. 5419 (1999), pp. 1,495–1,503; Mikhail A. Kreslavsky and James W. Head III, "Kilometer-Scale Roughness of Mars: Results from MOLA Data Analysis," *Journal of Geophysical Research: Planets,* 105, no. E11 (2000), pp. 26,695–26,711.

115 TILT SLIGHTLY Smith, et al., "The Global Topography of Mars and Implications for Surface Evolution," *Science.*

115 POSSIBLE SHORELINE J. Taylor Perron, et al., "Evidence for an Ancient Martian Ocean in the Topography of Deformed Shorelines," *Nature,* 447 (2007), pp. 840–843.

116 THE GROUND REBOUNDING Javier Ruiz, "On Ancient Shorelines and Heat Flows on Mars," Lunar and Planetary Science Conference, 36 (2005).

116 NOT JUST TOPOGRAPHY For instance, swaths of the planet's crust lit up with magnetic signals, just like an invisible-ink coloring book, indicating Mars had once been shielded from the solar wind, the deleterious gusts of particles from the sun that have been sputtering away its atmosphere ever since. There were minerals associated with chemical weathering, a clear signpost that the surface had interacted with water. There were also localized areas where Maria's altimeter suggested water could have pooled and formed ponds and evidence of gullies on slopes that didn't get much sunlight—in craters and valley walls, in alcoves spreading into fan-shaped aprons of debris. The spacecraft also sensed the presence of minuscule amounts of water in the atmosphere, and it peered out beyond the visible spectrum, out into the infrared, where minerals absorb particular wavelengths of light. Yet there was no evidence for carbonates, the chalky minerals that form in the presence of water and carbon dioxide. Our

planet's early carbon dioxide atmosphere—a whooping seventy bars of carbon dioxide, seventy times our atmospheric pressure—was pulled in the ocean, blanketing the floors of our ancient seas with limestone. Without a thicker carbon dioxide atmosphere, Mars couldn't have had stable surface water. That water should have led to carbonates, yet there were no carbonates to be found, and no one knew why.

116 WEATHER REPORTS Zuber, personal interview by Johnson; "Mars Orbiter Camera Mars Weather Update, for the week September 3–9, 2002," Malin Space Science Systems.

116 MARIA EVEN HEARD Zuber, personal interview by Johnson.

116 FIRST PHOTOGRAPH "PIA04531: Earth and Moon as Viewed from Mars," *Mars Global Surveyor* (May 22, 2003).

116 SEEING OUR PLANET Victoria Jaggard, "What Yuri Gagarin Saw on First Space Flight." *National Geographic*, April 13, 2011; "I see Earth! It is so beautiful!" European Space Agency (March 29, 2011).

116 "A MOTE OF DUST" Carl Sagan, *Pale Blue Dot: A Vision of the Human Future in Space* (New York: Ballantine Books, 1997), p. 6.

117 QUOTE FROM TOLKIEN J. R. R. Tolkien, *The Fellowship of the Ring* (New York: Del Rey, 1986), p. 193.

Chapter 8: The Acid Flats

119 JOHN GROTZINGER John is currently the Fletcher Jones Professor of Geology at Caltech and is the principal investigator of NASA's Mars Science Laboratory mission.

119 STARED LIKE A WOLF This is based on a remark by Paul Hoffman, an emeritus Harvard professor, who once described Grotzinger as having "eyes a bit like a wolf, when he looks at a rock." Quoted in Amina Khan, "Seeing Mars Through the Eyes of a Geologist," *www.phys.org* (Aug. 4, 2012).

119 WOULD BE BEAMED "The Rover's Antennas," NASA Mars; "Communications with Earth," NASA Mars.

119 IN NEARBY MONTROSE Daniel Siegal, "Montrose Jeweler Makes Watches on Mars Time," *Los Angeles Times* (Oct. 30, 2013).

120 ON THE FOURTH Steve Squyres, *Roving Mars: Spirit, Opportunity, and the Exploration of the Red Planet* (New York: Hyperion, 2005), p. 230.

120 WARM AND WET PAST "Mars Exploration Rovers Overview," *Mars Exploration Rovers*, NASA.

120 SIZE OF CONNECTICUT Guy Webster, "Go to That Crater and Turn Right: Spirit Gets a Travel Itinerary," NASA (January 13, 2004).

120 "BASALT PRISON" This nickname became a popular way for the team to describe the landing site at Gusev.

120 ONE OF THE SAFEST PLACES TO LAND Steven W. Squyres and Andrew H. Knoll, "Sedimentary Rocks at Meridiani Planum: Origin, Diagenesis, and Im-

plications for Life on Mars," *Earth and Planetary Science Letters* 240, no. 1 (2005), pp. 1–10.

120 CRYSTALLINE FORM OF RUST Bruce Murray, quoted in Jennifer Vaughn, "Mars, Old and New: A Personal View by Bruce Murray," The Planetary Society (September 3, 2013). Hematite (Fe2O3) appears red in its fine-grained form and gray in its coarse-grained form. Signatures of coarse-grained gray hematite in orbital infrared spectrometry data were what drove the decision to land at Merdiani Planum.

121 BEGAN CLAPPING The recounting of this moment, as well as many others in the early days of the Mars Exploration Rover mission, are encapsulated in Steve Squyres's wonderful firsthand account: Squyres, *Roving Mars: Spirit, Opportunity, and the Exploration of the Red Planet,* p. 292.

121 FELT COMPLETELY DISORIENTED Ibid., p. 292.

121 "WELCOME TO MERIDIANI" Ibid., p. 293.

121 ADVOCATED FOR MERIDIANI Ibid., p. 307.

121 "HOLY SMOKES" Ibid., pp. 293–294; misunderstanding the expression, Korea's major afternoon daily ran with the headline, THE SECOND MARS ROVER LANDS, SEEING MYSTERIOUS SMOKE.

122 "HOLE IN ONE" Squyres, *Roving Mars: Spirit, Opportunity, and the Exploration of the Red Planet,* p. 294.

122 BARELY ANKLE-HIGH Marcus Y. Woo, "Roving on Mars," Engineering and Science, 72 (2), (2009) pp. 12–20.

122 WERE DUBBED "BLUEBERRIES" "Martian 'Blueberries,'" NASA Science Mars Exploration Program (Jan. 27, 2015).

122 "FREAKY LITTLE HEMATITE BALLS" Squyres, *Roving Mars: Spirit, Opportunity, and the Exploration of the Red Planet,* p. 300.

123 SUGGESTING THAT THE HEMATITE HAD FORMED Marjorie A Chan, Brenda Beitler, W. T. Parry, Jens Ormö, and Goro Komatsu, "A Possible Terrestrial Analogue for Haematite Concretions on Mars," *Nature,* 429, no. 6993 (2004); Scott M. McLennan, J. F. Bell III, W. M. Calvin, P. R. Christensen, BC D. Clark, P. A. De Souza, J. Farmer et al., "Provenance and Diagenesis of the Evaporite-Bearing Burns Formation, Meridiani Planum, Mars," *Earth and Planetary Science Letters,* 240, no. 1 (2005), pp. 95–121; W. M. Calvin, et al., "Hematite Spherules at Meridiani: Results from MI, Mini-TES, and Pancam," *Journal of Geophysical Research: Planets,* 113, no. E12 (2008).

123 MAGNESIUM SULFATE EVERYWHERE Henry Bortman, "Evidence of Water Found on Mars," *Astrobiology Magazine* (March 3, 2004).

123 SULFATE MINERAL JAROSITE G. Klingelhöfer, R. Van Morris, B. Bernhardt, C. Schröder, D. S. Rodionov, P. A. De Souza, A. Yen, et al., "Jarosite and Hematite at Meridiani Planum from Opportunity's Mössbauer Spectrometer," *Science,* 306, no. 5702 (2004), pp. 1,740–1,745.

123 HIGHLY ACIDIC CONDITIONS This would have been a welcome development to Roger Burns, an MIT geologist, who proposed a model of acid weathering of

the iron-rich basalts on Mars long before anyone else; see: Roger Burns and Duncan Fisher, "Rates of Oxidative Weathering on the Surface of Mars," *Journal of Geophysical Research,* 98 (1993). For this reason, the evaporite-bearing Burns formation at Meridiani was named in his honor. Acidic waters also help explain the mystery of why so few chalky carbonates had been detected on the planet's surface, as carbonates do not precipitate out of solution at low pH.

123 "RIVER OF FIRE" Linda A. Amaral Zettler, Felipe Gómez, Erik Zettler, Brendan G. Keenan, Ricardo Amils, and Mitchell L. Sogin, "Microbiology: Eukaryotic Diversity in Spain's River of Fire," *Nature,* 417, no. 6885 (2002), p. 137.

123 LAYERS OF ROCK OVERLAPPING J. P. Grotzinger, R. E. Arvidson, J. F. Bell III, W. Calvin, B. C. Clark, D. A. Fike, M. Golombek, et al., "Stratigraphy and Sedimentology of a Dry to Wet Eolian Depositional System, Burns Formation, Meridiani Planum, Mars," *Earth and Planetary Science Letters,* 240, no. 1 (2005), pp. 11–72; McLennan, et al., "Provenance and Diagenesis of the Evaporite-Bearing Burns Formation, Meridiani Planum, Mars," *Earth and Planetary Science Letters.*

124 LIGHTHEARTEDNESS INSTANTLY RETURNED The team was broken into scientific areas of expertise—one for geology, one for atmospheric science, etc.—not into teams based on the rover's seven instruments. As I'd later realize, that was a deliberate decision made to dissolve rivalries among those who'd worked in small groups for years to build a particular piece of hardware. Instead of competing for time and power to be dedicated to their beloved technology during SOWG, the theme groups had to argue the merits of the larger scientific goals.

125 CRACK NAMED ANATOLIA "Press Release Images: Opportunity: A Puzzling Crack," NASA Mars Exploration Rovers (April 6, 2004).

125 CRATER NAMED FRAM S. W. Squyres, et al., "Overview of the Opportunity Mars Exploration Rover Mission to Meridiani Planum: Eagle Crater to Purgatory Ripple," *Journal of Geophysical Research,* 111 (2006), p. 4.

125 THE ROUTE TO ENDURANCE The convention soon became to name craters after famous ships of exploration: *Fram* was the ship that transported Amundsen's crew to Antarctica, where they first successfully reached the South Pole; *Endurance* was the name of Shackleton's ship that was crushed in the ice; and *Endeavor* was the name of the ship that took James Cook on his voyage to New Zealand and Australia. The team joked that if *Opportunity* ever reached Endeavor Crater, only a few graduate students would remain, just as only a few crew members managed to avoid malaria and dysentery on Cook's voyage. Even *Eagle* fit the convention, as it was also the name of the famous spacecraft that delivered Neil Armstrong and Buzz Aldrin to the surface of the moon.

125 "FIRST NAVCAM FRAME" Squyres, *Roving Mars: Spirit, Opportunity, and the Exploration of the Red Planet,* p. 335.

125 DRIVEN AND DRIVEN Ibid., p. 334.

126 A NEW MAP Bruce C. Heezen and Marie Tharp, "World Ocean Floor Panorama," full color, painted by H. Berann, Mercator projection, scale 1, no. 23,230,300 (1977).

127 SAGAN BARELY SET FOOT Steve went on to work closely with other Cornell professors, including Joseph Veverka (his scientific advisor), Arthur Bloom,

Steven Ostro, and William Travers, as well as Gene Shoemaker at USGS and several members of the *Voyager* imaging team. Steven Squyres, "The Morphology and Evolution of Ganymede and Callisto," Cornell PhD thesis (1981).

128 STIPPLED WITH BLUEBERRIES David R. Williams, "Mars Rover 'Opportunity' Images," NASA Goddard Space Flight Center (June 16, 2004).

128 COLUMBIA HILLS The Columbia Hills were named after the space shuttle *Columbia,* which disintegrated as it reentered Earth's atmosphere in 2003, killing the crew. The seven peaks are named for the seven individual astronauts.

128 WEST SPUR OF HUSBAND HILL Squyres, *Roving Mars: Spirit, Opportunity, and the Exploration of the Red Planet,* pp. 351–354.

128 UNCOVERED TRACES OF HEMATITE Squyres, *Roving Mars: Spirit, Opportunity, and the Exploration of the Red Planet,* pp. 351–354, 362–363.

128 CHEMICAL ENRICHMENTS Douglas Wayne Ming, David W. Mittlefehldt, Richard Van Morris, D. C. Golden, Ralf Gellert, Albert Yen, Benton C. Clark, et al., "Geochemical and Mineralogical Indicators for Aqueous Processes in the Columbia Hills of Gusev Crater, Mars," *Journal of Geophysical Research: Planets,* 111, no. E2 (2006).

128 WERE CALLED "VUGS" Squyres, Steven W., John P. Grotzinger, Raymond E. Arvidson, J. F. Bell, Wendy Calvin, Philip R. Christensen, Benton C. Clark, et al., "In Situ Evidence for an Ancient Aqueous Environment at Meridiani Planum, Mars," *Science,* 306, no. 5702 (2004), pp. 1,709–1,714; Kenneth E. Herkenhoff, S. W. Squyres, R. Arvidson, D. S. Bass, J. F. Bell, Pernille Bertelsen, B. L. Ehlmann, et al., "Evidence from Opportunity's Microscopic Imager for Water on Meridiani Planum," *Science,* 306, no. 5702 (2004), pp. 1,727–1,730.

129 LAST FEATURES TO FORM McLennan, et al, "Provenance and Diagenesis of the Evaporite-Bearing Burns Formation, Meridiani Planum, Mars," *Earth and Planetary Science Letters.*

129 WATER TABLE HAD RISEN S. W. Squyres and Andrew H. Knoll, *Sedimentary Geology at Meridiani Planum, Mars* (Houston: Gulf Professional Publishing, 2005), p. 68; J. P. Grotzinger, "Depositional Model for the Burns Formation, Meridiani Planum," *Seventh International Conference on Mars* (2007).

131 SEE OUR TRACKS NASA/JPL/MSSS, "Opportunity Tracks Seen From Orbit," NASA Science Mars Exploration Program (Jan. 24, 2005).

131 "FOLLOW THE WATER" "Fourth Planet from the Sun," NASA Mars Exploration Program.

131 JAROSITE THAT *OPPORTUNITY* HAD FOUND The Mössbauer spectrometer on the rover's arm was designed specifically to look at different types of iron minerals, which could help tease apart past environmental conditions, given that iron rusts in the presence of water and oxygen.

131 HIGHLY ACIDIC WATER Peter Cogram, "Jarosite," in *Reference Module in Earth Systems and Environmental Sciences,* Scott A. Elias, et al., eds. (*ScienceDirect,* 2018). While high concentrations of potassium ions, for instance, can also lead to jarosite formation at higher pH, the surface chemistry supported an acidic interpretation.

132 EVEN EUKARYOTES Amaral Zettler, Gómez, Zettler, Keenan, Amils, and Sogin, "Microbiology: Eukaryotic Diversity in Spain's River of Fire," *Nature.*

132 UNDER WAY FOR *SCIENCE* Nicholas J. Tosca, Andrew H. Knoll, and Scott M. McLennan, "Water Activity and Challenge for Life on Early Mars," *Science,* 320, no. 5880 (2008), pp. 1,204–1,207.

132 SOY SAUCE J. E. Henney, C. L. Taylor, and C. S. Boon, eds., "Preservation and Physical Property Roles of Sodium in Foods," in *Strategies to Reduce Sodium Intake in the United States* (Washington, D.C.: National Academies Press, 2010).

132 CALLED THE CLUSTERS "DAISIES" "Daisy Found on 'Route 66'," Mars Exploration Rovers *Spirit* Press Release Image, NASA/JPL/Cornell (April 17, 2004).

133 IT REMINDED ME OF Joseph von Littrow's proposed fiery canals in the Sahara, Carl Friedrich Gauss's triangle of wheat in Siberia, and Charles Cros's mirrors across Europe have been cited more than a dozen times in recent years, though it is not clear whether these proposals were genuine, speculative, or mere rumors. That these stories circulated in the eighteenth and nineteenth centuries, however, points to a preoccupation during that era with signaling to extraterrestrial beings that the Earth was inhabited by intelligent life-forms (in 1900, a prize of 100,000 francs—the "Prix Pierre Guzman"—was even set up by the French Académie des Sciences for the first person to communicate with a celestial object other than Mars—Mars was excluded because it was considered to be sufficiently well known, and therefore not a difficult enough challenge). Jeff Greenwald, "Who's Out There?" *Discover* (April 1, 1999); Michael J. Crowe, *The Extraterrestrial Life Debate, 1750–1900* (Mineola, N.Y.: Dover Publications, 2011), p. 205; Hans Zappe, *Fundamentals of Micro-Optics,* 1st ed. (Cambridge University Press, 2010), p. 298; Willy Ley, *Rockets, Missiles, and Space Travel* (New York: Viking Press, 1958); Frank Drake, "A Brief History of SETI," *Third Decennial US–USSR Conference on SETI—ASP Conference Series,* 47 (1993), pp. 11–18; Michael Carroll, *Earths of Distant Suns* (Göttingen, Germany: Copernicus), pp. 14–15; *Comptes Rendus Hebdomadaires des Séances de l'Académie des Sciences,* 131 (1900), p. 1,147.

133 CAVE OF SWIMMERS Gilf Kebir, László E. de Almásy, *Récentes Explorations dans le Désert Libyque (1932–1936),* E. and R. Schindler pour la Société Royale de Géographie d'Égypte, 1936.

Chapter 9: In Aeternum

137 "ROAD OF BONES" Marcus Warren, "'Road of Bones' Where Slaves Perished," *The Telegraph* (London: Aug. 10, 2002).

137 GATHER BLUEBERRIES Quoted in Kristofor Minta and Herbert Pföstl, *To Die No More* (New York: Blind Pony Books, 2008), originally from Catherine Merridale, *Night of Stone: Death and Memory in Twentieth Century Russia* (New York: Penguin Books, 2002), p. 300.

137 IN THE 1990S Eske Willerslev, et al., "Diverse Plant and Animal Genetic Records from Holocene and Pleistocene Sediments," *Science,* 300, no. 5,620 (2003), pp. 791–795 (see also "Sample Information" and "Stratigraphic Information" in "Supporting Material").

138 DICE-SIZED CUBES Carl Zimmer, "Eske Willerslev Is Rewriting History with DNA," *The New York Times* (May 16, 2016).

138 FROZEN WATER I. Mitrofanov, et al., "Maps of Subsurface Hydrogen from the High Energy Neutron Detector, Mars Odyssey," *Science*, 297, no. 5,578 (2002), pp. 78–81.

138 IN 2001 BY A NASA ORBITER This was NASA's *Mars Odyssey* mission, which launched in April of 2001 and arrived that October.

139 NEWLY FORMED GROUP Eske continues to do pioneering work in evolutionary genetics; he is now at the University of Copenhagen, where he leads the Center for GeoGenetics.

139 ONCE DISAPPEARED Zimmer, "Eske Willerslev Is Rewriting History with DNA," *The New York Times*.

139 CONTAMINATION WAS SUCH Most genomics labs were awash in tiny strands of DNA—exponentially copied target molecules that would waft out of strips of plastic tubes as they clicked open. To work with such lean samples—where there was low signal to noise—we needed to transfer our sealed DNA extractions to another building before we amplified them and began our analyses.

140 SØREN KIERKEGAARD'S GRAVE Joakim Garff, translated by Bruce H. Kirmmse, *Søren Kierkegaard: A Biography* (Princeton, N.J.: Princeton University Press, 2000), p. 811.

140 300,000 YEARS Sarah S. Johnson, et al., "Ancient Bacteria Show Evidence of DNA Repair," *Proceedings of the National Academy of Sciences*, 104 (36) (2007), pp. 14,401–14,405.

143 THE ENTIRE PHYLOGENIC TREE Rod Pyle, *Destination Mars: New Explorations of the Red Planet* (Amherst, N.Y.: Prometheus Books, 2012), p. 248.

143 ASSESS ITS BIOLOGICAL POTENTIAL The official objectives of the mission were to 1) study the history of water in the Martian arctic and 2) search for evidence of a habitable zone and assess the biological potential of the ice-soil boundary. "Mars Phoenix Lander Overview," NASA.

143 MYTHICAL ARABIAN BIRD Herodotus, *Histories*, trans. George Rawlinson (The Internet Classics Archive), II.

143 RECYCLED HARDWARE AND SOFTWARE Pyle, *Destination Mars: New Explorations of the Red Planet*, p. 231.

143 SHOESTRING BUDGET "NASA's Phoenix Mars Mission Gets Thumbs Up for 2007 Launch," NASA press release (June 2, 2005).

143 FIRST *SCOUT* MISSION "Phoenix Mars Scout," NASA Facts, NASA JPL.

143 RUN AND OPERATED Pyle, *Destination Mars: New Explorations of the Red Planet*, p. 230.

143 PULSE THRUSTERS D. H. Plemmons, et al., "Effects of the Phoenix Lander Descent Thruster Plume on the Martian Surface," *Journal of Geophysical Research*, 113 (2008).

143 ALMOST ALL THE SUCCESSFUL MISSIONS An exception includes the *Viking 2* lander, which touched down at 48 degrees north latitude.

143 CANADA'S NORTHWEST TERRITORIES "Frequently Asked Questions," Phoenix Mars Mission, the University of Arizona.

143 SLOW IT DOWN Eric Hand, "Mars exploration: Phoenix: a race against time," *Nature* (Dec. 10, 2008).

144 JUST BENEATH THE SURFACE W. C Feldman, W. V. Boynton, R. L. Tokar, T. H. Prettyman, O. Gasnault, S. W. Squyres, R. C. Elphic, et al., "Global Distribution of Neutrons from Mars: Results from Mars Odyssey," *Science,* 297, no. 5578 (2002), pp. 75–78; I. Mitrofanov, D. Anfimov, A. Kozyrev, M. Litvak, A. Sanin, V. Tret'yakov, A. Krylov, et al., "Maps of Subsurface Hydrogen from the High Energy Neutron Detector, Mars Odyssey." *Science,* 297, no. 5578 (2002), pp. 78–81.

144 A FIERY MURAL Angela Poulson, "UA Art Class About to Complete Giant Phoenix Mars Mission Mural," *UA News* (Dec. 1, 2006).

144 REHEARSING A PRESS CONFERENCE Pyle, *Destination Mars: New Explorations of the Red Planet,* p. 249.

144 VACCINE FOR YELLOW FEVER Joe Bargmann, "Spacemen," *The Washington Post Magazine* (Sept. 28, 2008).

144 A TWITTER ACCOUNT Alexis Madrigal, "Wired Science Scores Exclusive Twitter Interview with the Phoenix Mars Lander," *Wired* (May 30, 2008).

144 140-CHARACTER LIMIT Ibid.

144 "VEGAS SLOT MACHINE" Ibid.

144 *MARS RECONNAISSANCE ORBITER* The *Mars Reconnaissance Orbiter* was a NASA orbital mission that launched in 2005 and arrived in 2006. The mission has mapped the presence of clays, carbonates, and chlorides, determined the volume of water ice in the northern polar cap, and collected arrestingly detailed images of features like recurring slope lineae with its High Resolution Imaging Science Experiment (HiRISE) camera. The orbiter remains active today and is a key telecommunications link for rover surface operations.

144 PHOTOGRAPH OF *PHOENIX* John Mahoney, "Mars Reconnaissance Orbiter Captures Images of Phoenix Lander's Descent," *Popular Science* (May 27, 2008).

144 SIX AND A HALF SECONDS Ivan Semeniuk, "First Phoenix Images Reveal 'Quilted' Martian Terrain," *New Scientist* (May 26, 2008).

145 PIROUETTE AS IT OPENED Emily Lakdawalla, "Phoenix Has Landed!" The Planetary Society (May 25, 2008).

145 "CHEERS! TEARS!!" Madrigal, "Wired Science Scores Exclusive Twitter Interview with the Phoenix Mars Lander," *Wired.*

145 MULTIPLE CAMERAS ON *PHOENIX* Peter H. Smith, "Introduction to Visions of Mars," The Planetary Society (Feb. 14, 2007).

145 INTERSECTING POLYGONS Michael T. Mellon, Michael C. Malin, Raymond E. Arvidson, Mindi L. Searls, Hanna G. Sizemore, Tabatha L. Heet, Mark T. Lemmon, H. Uwe Keller, and John Marshall, "The Periglacial Landscape at the Phoenix Landing Site," *Journal of Geophysical Research: Planets,* 114, no. E1 (2009).

145 REPEATED EXPANSION AND CONTRACTION Ivan Semeniuk, "First Phoenix Images Reveal 'Quilted' Martian Terrain," *New Scientist* (May 26, 2008).

145 "POLYGONS WITHIN POLYGONS WITHIN POLYGONS" Ivan Semeniuk, "Mars Scientists Ponder Polygon Mystery," *New Scientist* (May 27, 2008).

145 FROM THE UNIVERSITY OF MICHIGAN That team member was Nilton Rennó, a professor of climate and space sciences and engineering at the University of Michigan.

145 WATER DROPLETS Kenneth Chang, "Blobs in Photos of Mars Lander Stir a Debate: Are They Water?" *The New York Times* (March 16, 2009); N. Rennó, et al., Lunar and Planetary Science Conference, 40 (2009); Nilton O. Rennó, Brent J. Bos, David Catling, Benton C. Clark, Line Drube, David Fisher, Walter Goetz, et al., "Possible Physical and Thermodynamical Evidence for Liquid Water at the Phoenix Landing Site," *Journal of Geophysical Research: Planets,* 114, no. E1 (2009).

145 SMITH TOLD THE PRESS Andrea Thompson, "Phoenix Mars Lander Found Liquid Water, Some Scientists Think," *Space* (March 10, 2009).

145 ANOTHER WHITE PATCH P. H. Smith, L. K. Tamppari, R. E. Arvidson, D. Bass, D. Blaney, William V. Boynton, A. Carswell, et al., "H2O at the Phoenix Landing Site," *Science,* 325, no. 5936 (2009), pp. 58–61.

146 "IT MUST BE ICE" Guy Webster, "Bright Chunks at Phoenix Lander's Mars Site Must Have Been Ice," NASA (June 19, 2008).

146 TEGA AND MECA TEGA was the acronym for *Phoenix*'s Thermal and Evolved Gas Analyzer, and MECA was the acronym for *Phoenix*'s Microscopy, Electrochemistry, and Conductivity Analyzer.

146 DELIVERING THE SAMPLES PROVED Eric Hand, "Mars Exploration: Phoenix: A Race Against Time," *Nature* (Dec. 10, 2008).

146 CALCIUM CARBONATE W. V. Boynton, D. W. Ming, S. P. Kounaves, S. M. M. Young, R. E. Arvidson, M. H. Hecht, J. Hoffman, et al., "Evidence for Calcium Carbonate at the Mars Phoenix Landing Site," *Science,* 325, no. 5936 (2009), pp. 61–64.

146 WATERY PAST Unfortunately, there wasn't enough water to get an isotopic reading, which would have helped to understand how much water had been lost from the planet.

146 WET CHEMISTRY LAB "Microscopy, Electrochemistry, and Conductivity Analyzer (MECA)," Phoenix Mars Mission, University of Arizona.

146 SUGAR-CUBE-SIZED The soil samples were one cubic centimeter in volume. See: S. P. Kounaves, "The Phoenix Mars Lander Wet Chemistry Laboratory (WCL): Understanding the Aqueous Geochemistry of the Martian Soil," International Workshop on Instrumentation for Planetary Missions, vol. 1,683 (2012), p. 1005.

146 TINY SENSORS Samuel P. Kounaves, Michael H. Hecht, Steven J. West, John-Michael Morookian, Suzanne M. M. Young, Richard Quinn, Paula Grunthaner, et al., "The MECA Wet Chemistry Laboratory on the 2007 *Phoenix* Mars Scout Lander," *Journal of Geophysical Research: Planets,* 114, no. E3 (2009).

146 SLIGHTLY ALKALINE Elizabeth K. Wilson, "Mars Soil PH Measured," *Chemical and Engineering News* (June 27, 2008).

146 INCLUDING NITRATE Samuel P. Kounaves, et al., "Evidence of Martian Perchlorate, Chlorate, and Nitrate in Mars Meteorite EETA79001: Implications

for Oxidants and Organics," *Icarus,* 229 (2014), pp. 206–213. We assume nitrogen gas was released into the Martian atmosphere via the exhalations of ancient volcanoes, just as it was here on Earth. Even though life is swimming in nitrogen gas on Earth, the nitrogen isn't accessible to most organisms: The triple bond is simply too hard to break. Other forms of nitrogen, like nitrate, which hold to oxygen atoms with a bond that can be easily cleaved, are often necessary.

146 HYPOTHESIZED TO EXIST Rocco L. Mancinelli and Amos Banin, "Where Is the Nitrogen on Mars?" *International Journal of Astrobiology,* 2, no. 3 (2003), pp. 217–225.

146 DESIGNED TO DETECT NITRATE It would be another few years before nitrate was definitively detected on Mars, by NASA Goddard scientist Jen Stern, using the Sample Analysis at Mars instrument aboard *Curiosity*. Jennifer C. Stern, Brad Sutter, Caroline Freissinet, Rafael Navarro-González, Christopher P. McKay, P. Douglas Archer, Arnaud Buch, et al., "Evidence for Indigenous Nitrogen in Sedimentary and Aeolian Deposits from the Curiosity Rover Investigations at Gale Crater, Mars," *Proceedings of the National Academy of Sciences,* 112, no. 14 (2015), pp. 4,245–4,250.

147 SALT CALLED PERCHLORATE Perchlorate is a molecule containing a perchlorate ion (ClO4-, a chlorine atom bonded to four oxygen atoms). The discovery of perchlorate at the Phoenix landing site is detailed here: M. H. Hecht, S. P. Kounaves, R. C. Quinn, S. J. West, S. M. M. Young, D. W. Ming, D. C. Catling, et al., "Detection of Perchlorate and the Soluble Chemistry of Martian Soil at the Phoenix Lander Site," *Science,* 325, no. 5936 (2009), pp. 64–67.

147 LOOK IT UP Leonard David, "Toxic Mars: Astronauts Must Deal with Perchlorate on the Red Planet," *Space* (June 13, 2013).

147 JUST A FEW HANDFULS Quoting Mike Hecht, see: Ryan Anderson, "AGU Day 1: Phoenix," AGU 100 Blogosphere (Dec. 16, 2008).

147 CHLORINATED MOLECULES Chloromethane and dichloromethane, e.g. See: Rafael Navarro-González, Edgar Vargas, Jose de La Rosa, Alejandro C. Raga, and Christopher P. McKay, "Reanalysis of the Viking Results Suggests Perchlorate and Organics at Midlatitudes on Mars," *Journal of Geophysical Research: Planets,* 115, no. E12 (2010).

147 RESIDUES OF CLEANING FLUIDS Guy Webster, "Missing Piece Inspires New Look at Mars Puzzle," NASA (Sept. 3, 2010).

147 WOULD HAVE DESTROYED Kounaves, "The Phoenix Mars Lander Wet Chemistry Laboratory (WCL): Understanding the Aqueous Geochemistry of the Martian Soil," International Workshop on Instrumentation for Planetary Missions, p. 1005.

147 NECESSARILY BAD FOR MICROBES See, for example: John D. Coates and Laurie A. Achenbach, "Microbial Perchlorate Reduction: Rocket-Fueled Metabolism," *Nature Reviews Microbiology,* 2, no. 7 (2004), p. 569; Joop M. Houtkooper and Dirk Schulze-Makuch, "The Possible Role of Perchlorates for Martian Life," *Journal of Cosmology,* Vol. 5 (Jan. 25, 2010), pp. 930–939; Sophie Nixon, Claire Rachel Cousins, and Charles Cockell, "Plausible Microbial Metabolisms on Mars," *Astronomy & Geophysics* (2013).

147 THE FREEZING POINT Theoretical eutectic values were determined be 236 ± 1 K for 52 wt% sodium perchlorate and 206 ± 1 K for 44.0 wt% magnesium perchlorate; see: Vincent F. Chevrier, Jennifer Hanley, and Travis S. Altheide, "Stability of Perchlorate Hydrates and Their Liquid Solutions at the Phoenix Landing Site, Mars," *Geophysical Research Letters,* 36, no. 10 (2009).

148 "A BLACKBOARD HERE" Mars *Phoenix,* Twitter post (July 8, 2008, 3:15 P.M.).

148 "FINISHED HUNKERING DOWN" Ibid. (Oct. 8, 2008, 10:20 P.M.).

148 TWEETS TURNED REFLECTIVE Ryan Anderson, "Phoenix Hanging in There," AGU 100 Blogosphere (Oct. 31, 2008).

148 "HEATER TURNING OFF" Mars *Phoenix,* Twitter post (Oct. 28, 2008, 4:55 P.M.).

148 LIKE CIRRUS CLOUDS J. A. Whiteway, et al., "Mars Water-Ice Clouds and Precipitation," *Science,* 325, no. 5,936 (2009), pp. 68–70.

148 "BARELY WETTING THE SURFACE" Anne Minard, "'Diamond Dust' Snow Falls Nightly on Mars," *National Geographic News* (July 2, 2009).

148 "LAZARUS MODE" Eric Hand, "Mars exploration: Phoenix: a race against time," *Nature* (Dec. 10, 2008).

148 "01010100 01110010 01101001" Mars *Phoenix,* Twitter post (Nov. 10, 2008, 1:12 P.M.); Rod Pyle, *Destination Mars: New Explorations of the Red Planet* (Amherst, N.Y.: Prometheus Books, 2012), p. 241.

149 A MINI-DVD "Visions of Mars," The Planetary Society.

149 VELCRO'D TO THE DECK Bruce Betts, "We Make It Happen," *The Planetary Report,* vol. XXVI, no. 6 (2006), p. 3.

149 RUSSIAN MARS MISSION Jon Lomberg, "Visions of Mars: Then and Now," The Planetary Society.

149 "ATTENTION ASTRONAUTS" Ibid.

149 BOOKS AND STORIES "Visions of Mars: The Stories," The Planetary Society.

149 SURREALIST PAINTINGS "Visions of Mars: Artwork and Radio Broadcasts," The Planetary Society.

149 THE 1940 RECORDING Ibid.

150 IN HIS INTRODUCTION Peter H. Smith, "Introduction to Visions of Mars," The Planetary Society (Feb. 14, 2007).

150 "WOULD NOT BE CURRENT" Ibid.

150 HUNDREDS OF YEARS The mini-DVD is expected to last approximately five hundred years. Bruce Betts, "We Make It Happen," *The Planetary Report.*

150 THE OLDEST PIECES INCLUDED "Visions of Mars: The Stories," The Planetary Society.

150 VOLTAIRE'S "MICROMÉGAS" Voltaire, "Micromégas," Blake Linton Wilfong, ed., Free Sci-Fi Classics.

151 STUDY *DROSOPHILA* Jeffrey R. Powell, *Progress and Prospects in Evolutionary Biology: The Drosophila Model* (Oxford University Press, 1997).

152 DUTCH BOOK Kees Boeke, *Cosmic View: The Universe in 40 Jumps* (New York: John Day Company, 1957). This book was also the inspiration for a famous short film, *Powers of Ten,* produced by Ray and Charles Eames in 1977.

Chapter 10: Sweet Water

155 POOLED AND FORMED LAKES "Mars Science Laboratory Landing Site: Gale Crater," NASA Mars Exploration Program (July 22, 2011).

156 INTO THE DEEP NIGHT "NASA Launches Most Capable and Robust Rover to Mars," NASA Mars Exploration Program (Nov. 26, 2011).

156 LANDING VIA A "SKY CRANE" Ravi Prakash, P. Dan Burkhart, Allen Chen, Keith A. Comeaux, Carl S. Guernsey, Devin M. Kipp, Leila V. Lorenzoni, et al., "Mars Science Laboratory Entry, Descent, and Landing System Overview," IEEE Aerospace Conference (2008), pp. 1–18.

158 FEEL A LITTLE CLOSER "Raw Images: Sol 3," Mars *Curiosity* Rover Raw Images (Aug, 8, 2012).

159 MOUNT SHARP Officially named Aeolis Mons by the International Astronomical Union.

159 THICKEST GEOLOGIC RECORDS "Press Kit, Mars Science Laboratory Landing," NASA (July 2012). Like "the most complete copy of an ancient manuscript," Mount Sharp can be used to help piece together less complete geologic records from other parts of the planet.

159 FILLED WITH SOFT CLAYS R. E. Milliken, J. P. Grotzinger, and B. J. Thomson, "Paleoclimate of Mars as Captured by the Stratigraphic Record in Gale Crater," *Geophysical Research Letters,* 37, no. 4 (2010); A. A. Fraeman, et al., "The stratigraphy and Evolution of Lower Mount Sharp from Spectral, Morphological, and Thermophysical Orbital Data Sets," *Journal of Geophysical Research: Planets,* 121 (2016), pp. 1,713–1,736.

159 "FOLLOW THE WATER" NASA's goals for Mars have progressed from "Follow the Water" (with several previous missions) to "Explore Habitability" (with NASA's *Curiosity* rover) to "Seek Signs of Life" (with NASA's *Mars 2020* mission). See: "The Mars Exploration Program," NASA Mars Exploration.

159 JUST NORTH "Context of Curiosity Landing Site in Gale Crater," NASA Mars Exploration Program (July 22, 2011).

159 FACING EAST-SOUTHEAST Emily Lakdawalla, "Curiosity: Notes from the Two Day-after-Landing Press Briefings," The Planetary Society (Aug. 6, 2012).

159 SETTING THE ROVER DOWN Jason Hanna, "'Impressive' Curiosity Landing Only 1.5 Miles Off, NASA Says," CNN (Aug. 14, 2012).

159 A FEW KILOMETERS AWAY *Curiosity* would first need to navigate around the Bagnold Dunes.

159 FIRST HUMAN VOICE Charles Bolden, quoted in "First Recorded Voice from Mars," Mars Science Laboratory, NASA (Aug. 27, 2012).

159 TEST ITS COMPONENTS Emily Lakdawalla, "Curiosity Sol 9 Update," The Planetary Society (Aug. 15, 2012).

161 AN ANCIENT RIVERBED R. M. E. Williams, J. P. Grotzinger, W. E. Dietrich, S. Gupta, D. Y. Sumner, R. C. Wiens, N. Mangold, et al., "Martian Fluvial Conglomerates at Gale Crater," *Science,* 340, no. 6,136 (2013), pp. 1,068–1,072.

161 PERHAPS AS DEEP AS MY HIP Guy Webster, "NASA Rover Finds Old Streambed on Martian Surface," NASA (September 27, 2012).

161 THREE DIFFERENT TYPES J. P. Grotzinger, D. Y. Sumner, L. C. Kah, K. Stack, S. Gupta, L. Edgar, D. Rubin, et al., "A Habitable Fluvio-Lacustrine Environment at Yellowknife Bay, Gale Crater, Mars," *Science,* 343, no. 6,169 (2014), p. 1,242,777.

161 EXPANSIVE LAKE Ibid.; J. P. Grotzinger, S. Gupta, M. C. Malin, D. M. Rubin, J. Schieber, K. Siebach, D. Y. Sumner, et al., "Deposition, Exhumation, and Paleoclimate of an Ancient Lake Deposit, Gale Crater, Mars," *Science,* 350, no. 6,257 (2015), p. aac7575; J. A. Hurowitz, J. P. Grotzinger, W. W. Fischer, S. M. McLennan, R. E. Milliken, N. Stein, A. R. Vasavada, et al., "Redox Stratification of an Ancient Lake in Gale Crater, Mars," *Science,* 356, no. 6341 (2017), p. eaah6849; C. Freissinet, D. P. Glavin, Paul R. Mahaffy, K. E. Miller, J. L. Eigenbrode, R. E. Summons, A. E. Brunner, et al., "Organic Molecules in the Sheepbed Mudstone, Gale Crater, Mars," *Journal of Geophysical Research: Planets,* 120, no. 3 (2015), pp. 495–514.

161 STOOD AS AN ISLAND "A Guide to Gale Crater," NASA Video (Aug. 2, 2017).

162 PAUL MAHAFFY Paul is now the director of the Solar System Exploration Division at NASA Goddard.

162 SAMPLE ANALYSIS AT MARS Paul R. Mahaffy, et al., "The Sample Analysis at Mars Investigation and Instrument Suite," *Space Science Reviews,* 170, no. 1–4 (2012), pp. 401–478.

162 WHEN HE WASN'T STUDYING Salem Solomon, "Born and Raised in Senafe, Eritrea, a NASA Scientist Leads Missions in Space," *Eritrean Press* (Aug. 20, 2015).

162 GREAT IRON CROSS Ibid.

162 LISTENING TO THE WIND Ibid.

162 WEIGHING NEARLY AS MUCH SAM comprised 42 percent of the combined instrument weight.

162 ANNOUNCED THE DISCOVERY Walter Sullivan, "Two Gases Associated with Life Found on Mars Near Polar Cap," *The New York Times* (Aug. 8, 1969).

162 A MONTH LATER Johnny Bontemps, "Mystery Methane on Mars: The Saga Continues," *Astrobiology Magazine* (May 14, 2015). It turned out that the absorptions over the polar cap could be explained by CO_2 ice—see: K. C. Herr and G. C. Pimentel, "Infrared Absorptions Near 3 Microns Recorded over Polar Cap of Mars," *Science,* 166 (1969), pp. 496–499.

162 CHILE AND HAWAII REPORTED M. J. Mumma, R. E. Novak, M. A. DiSanti, B. P. Bonev, and N. Dello Russo, "Detection and Mapping of Methane and Water on Mars," *Bulletin of the American Astronomical Society,* vol. 36 (2004), p. 1,127; Vladimir A. Krasnopolsky, Jean Pierre Maillard, and Tobias C. Owen, "Detection of Methane in the Martian Atmosphere: Evidence for Life?" *Icarus,* 172, no. 2 (2004), pp. 537–547; Michael J. Mumma, Geronimo L. Villanueva, Robert E. Novak, Tilak Hewagama, Boncho P. Bonev, Michael A. DiSanti, Avi M. Man-

dell, and Michael D. Smith, "Strong Release of Methane on Mars in Northern Summer 2003," *Science*, 323, no. 5,917 (2009), pp. 1,041–1,045.

162 ALSO PICKED UP TRACES Vittorio Formisano, Sushil Atreya, Thérèse Encrenaz, Nikolai Ignatiev, and Marco Giuranna, "Detection of Methane in the Atmosphere of Mars," *Science*, 306, no. 5,702 (2004), pp. 1,758–1,761.

162 A SEASONAL PATTERN Christopher R. Webster, Paul R. Mahaffy, Sushil K. Atreya, Gregory J. Flesch, Michael A. Mischna, Pierre-Yves Meslin, Kenneth A. Farley, et al., "Mars Methane Detection and Variability at Gale Crater," *Science*, 347, no. 6,220 (2015), pp. 415–417; Christopher R. Webster, Paul R. Mahaffy, Sushil K. Atreya, John E. Moores, Gregory J. Flesch, Charles Malespin, Christopher P. McKay, et al., "Background Levels of Methane in Mars' Atmosphere Show Strong Seasonal Variations," *Science*, 360, no. 6,393 (2018), pp. 1,093–1,096.

162 GEOLOGICAL PROCESS James R. Lyons, Craig Manning, and Francis Nimmo, "Formation of Methane on Mars by Fluid-Rock Interaction in the Crust," *Geophysical Research Letters*, 32, no. 13 (2005).

162 MATRICES OF MELTING ICE Brendon K. Chastain and Vincent Chevrier, "Methane Clathrate Hydrates as a Potential Source for Martian Atmospheric Methane," *Planetary and Space Science*, 55, no. 10 (2007), pp. 1,246–1,256.

162 EXHALATIONS OF A SMALL Michael D. Max and Stephen M. Clifford, "The State, Potential Distribution, and Biological Implications of Methane in the Martian Crust," *Journal of Geophysical Research: Planets*, 105, no. E2 (2000), pp. 4,165–4,171.

162 SAM ALSO CARRIED Paul R. Mahaffy, Christopher R. Webster, Michel Cabane, Pamela G. Conrad, Patrice Coll, Sushil K. Atreya, Robert Arvey, et al., "The Sample Analysis at Mars Investigation and Instrument Suite," *Space Science Reviews*, 170, no. 1–4 (2012), pp. 401–478.

163 DECIDED TO EXHUME Michael Farquhar, "Remains to Be Seen," *The Washington Post* (June 30, 1991).

163 STICK AROUND FOR BILLIONS OF YEARS Roger E. Summons, Pierre Albrecht, Gene McDonald, and J. Michael Moldowan, "Molecular Biosignatures," *Strategies of Life Detection* (Boston: Springer, 2008), pp. 133–159.

163 FOUND NO DEFINITIVE EVIDENCE OF ORGANICS Two key papers on the *Viking* GC-MS results are: K. Biemann, et al., "Search for Organic and Volatile Inorganic Compounds in Two Surface Samples from the Chryse Planitia Region of Mars," *Science*, 194 (1976), pp. 72–76; K. Biemann, et al., "The Search for Organic Substances and Inorganic Volatile Compounds in the Surface of Mars," *J. Geophys. Res.*, 82 (28) (1977), pp. 4,641–4,658. These results we reinterpreted after the discovery of perchlorate on Mars; see: Rafael Navarro-González, Edgar Vargas, Jose de La Rosa, Alejandro C. Raga, and Christopher P. McKay, "Reanalysis of the Viking Results Suggests Perchlorate and Organics at Midlatitudes on Mars," *Journal of Geophysical Research: Planets*, 115, no. E12 (2010).

163 ONE PART PER BILLION "SAM," Mars *Curiosity* Rover, NASA.

164 A BABY ASPIRIN Daniel Limonadi, "Sampling Mars, Part 2: Science Instruments SAM and Chemin," The Planetary Society (Aug. 20, 2012).

164 NEUTRAL PH CONDITIONS "View Into 'John Klein' Drill Hole in Martian Mudstone," NASA.

164 "ABLE TO DRINK IT" "Quotation of the Day for Wednesday, Mar. 13, 2013," *The New York Times;* Carl Franzen, "Curiosity Discovers Ancient Mars Could Have Supported Life," *The Verge* (March 12, 2013).

164 LONG THIN TUBE Mahaffy et. al., "The Sample Analysis at Mars Investigation and Instrument Suite," *Space Science Reviews;* Emily Lakdawalla, *The Design and Engineering of Curiosity: How the Mars Rover Performs Its Job* (Cham, Switzerland: Springer, 2018).

164 SIMPLE COMPOUNDS C. Freissinet, et al., "Organic Molecules in the Sheepbed Mudstone, Gale Crater, Mars," *Journal of Geophysical Research: Planets.*

164 MORE-COMPLEX MOLECULES, BOUND TOGETHER BY SULFUR J. L. Eigenbrode, R. E. Summons, A. Steele, C. Freissinet, M. Millan, R. Navarro-González, B. Sutter, et al., "Organic Matter Preserved in 3-Billion-Year-Old Mudstones at Gale Crater, Mars," *Science,* 360, no. 6,393 (2018), pp. 1,096–1,101; see also: Dwayne Brown and JoAnna Wendel, "NASA Finds Ancient Organic Material, Mysterious Methane on Mars," NASA (June 7, 2018).

164 BUILDING BLOCKS OF LIFE Some of the most exciting results from SAM may be yet to come. The heating method used thus far, pyrolysis, breaks complex organic molecules into simpler components. But nine "wet chemistry" cups have been included as part of SAM's carousel. Within these cups are solvents that combine with organics. In the presence of these solvents, much less heating is required, lessening the combustion, enabling larger, more complex molecules to waft into SAM's GC-MS.

164 HOLES IN ITS WHEELS Mike Wall, "NASA's Curiosity Rover on Mars Is Climbing a Mountain Despite Wheel Damage," *Space* (May 3, 2016).

164 POOR DRILL *Curiosity*'s drill went on hiatus in late 2016 after a motor failed; a new percussive drilling technique (called Feed Extended Drilling) designed by JPL's engineers successfully brought the drill back to life eighteen months later in 2018; Guy Webster, "Curiosity Rover Team Examining New Drill Hiatus," NASA (Dec. 5, 2016); "Curiosity Successfully Drills 'Duluth'," NASA Science Mars Exploration Program (May 23, 2018).

164 VERA RUBIN RIDGE A special issue of the *Journal of Geophysical Research* detailing discoveries in the Vera Rubin Ridge is slated to be released in 2020.

164 EVIDENCE OF DARK MATTER Vera C. Rubin, W. Kent Ford, Jr., and Norbert Thonnard, "Rotational Properties of 21 SC Galaxies with a Large Range of Luminosities and Radii, from NGC 4605 (R= 4kpc) to UGC 2885 (R= 122 kpc)," *The Astrophysical Journal,* 238 (1980), pp. 471–487; J. G. De Swart, Gianfranco Bertone, and Jeroen van Dongen, "How Dark Matter Came to Matter," *Nature Astronomy,* 1, no. 3 (2017), p. 0059.

165 85 PERCENT OF THE "Synopsis: How Dark Matter Shaped the First Galaxies," American Physical Society (Oct. 2, 2019).

165 LEARNED TO TELL TIME Interview of Vera Rubin by Alan Lightman, Niels Bohr Library and Archives, American Institute of Physics (College Park, Md.: April 3, 1989).

165 PAINTING ASTRONOMICAL OBJECTS Maiken Scott, "Vera Rubin's Son Reflects on How She Paved the Way for Women," *The Pulse* (Jan. 12, 2017).

165 ASSISTANT PROFESSOR AT A TIME "Dark Matter Discoverer Vera Rubin Blazed New Trails for Women, Astronomy," Georgetown University News (Feb. 23, 2017).

165 NOT OPEN TO WOMEN In 1965, Rubin also became the first woman to observe at Caltech's Palomar Observatory. As Princeton professor Neta Bahcall remembers: "They told her, 'It's a real problem because we don't have a ladies' room,' so she went back to her room and took out a little piece of paper and cut it into a skirt and went to the bathroom door and stuck it on the men's figure on the door. She said, 'Look, now you have a ladies' room.' " See: Jenni Avins, " 'Devise Your Own Paths': The Enduring Wisdom of Vera Rubin, Groundbreaking Astronomer and Working Mother," *Quartz* (Dec. 27, 2016).

165 AVERSE TO SHARP ELBOWS Dennis Overbye, "Vera Rubin, 88, Dies; Opened Doors in Astronomy, and for Women," *The New York Times* (Dec. 27, 2016).

165 "NOBODY WOULD BOTHER [HER]" Ibid.; Lisa Randall, "Why Vera Rubin Deserved a Nobel," *The New York Times* (Jan. 4, 2017).

165 MAJORITY OF THE UNIVERSE Overbye, "Vera Rubin, 88, Dies; Opened Doors in Astronomy, and for Women," *The New York Times.*

165 NEAR THE END OF HER LIFE Rubin was only the second woman, after Caroline Herschel in 1828, to win the Royal Astronomical Society's Gold Medal. She also received a National Medal of Science in 1993, the highest scientific award in the United States. Many also believe she should have been awarded a Nobel Prize for her work; see: Randall, "Why Vera Rubin Deserved A Nobel," *The New York Times,* and Sarah Scoles, "How Vera Rubin Confirmed Dark Matter," *Astronomy* (June 2016).

165 "TO HAVE A FAMILY" Vera Rubin, *Bright Galaxies, Dark Matters* (New York, Springer Science and Business Media: 1996).

165 SHE HAD FOUR CHILDREN Overbye, "Vera Rubin, 88, Dies; Opened Doors in Astronomy, and for Women," *The New York Times.*

165 BECAME A PLANETARY GEOLOGIST Dave Rubin is a sedimentologist in the department of Earth and planetary sciences at U.C. Santa Cruz and a participating scientist on the Mars Science Laboratory science team, focusing on sediment deposits and geomorphology.

165 "I'M SORRY I KNOW SO LITTLE" Overbye, "Vera Rubin, 88, Dies; Opened Doors in Astronomy, and for Women," *The New York Times.*

Chapter 11: Form from a Formless Thing

167 A VILLAGE NAMED JEZERO D. C. Agle, "NASA Mars Mission Connects with Bosnian Town," NASA News (Sept. 23, 2019).

167 LINKED BY A HUNDRED WATERFALLS Plitvice Lakes National Park, Croatia. See: "Discover the Most Beautiful Lakes in Eastern Europe," SNCB International.

167 RUMORED TO BE HEAVY WITH DEUTERIUM "Pliva Lakes and Watermills," Visit Jajce.

167 NAMED AFTER SMALL TOWNS "Categories (Themes) for Naming Features on Planets and Satellites," *Gazetteer of Planetary Nomenclature,* International Astronomical Union.

167 SMALL CRATER NAMED "Jezero," *Gazetteer of Planetary Nomenclature,* International Astronomical Union.

167 ALSO HELD A LAKE C. I. Fassett and J. W. Head III, "Fluvial Sedimentary Deposits on Mars: Ancient Deltas in a Crater Lake in the Nili Fossae Region," *Geophysical Research Letters* 32, no. 14 (2005); B. L. Ehlmann, J. F. Mustard, C. I. Fassett, S. C. Schon, J. W. Head III, D. J. Des Marais, J. A. Grant, and S. L. Murchie, "Clay Minerals in Delta Deposits and Organic Preservation Potential on Mars," *Nature Geoscience* 1, no. 6 (2008): 355; T.A. Goudge, "Stratigraphy and Evolution of Delta Channel Deposits, Jezero Crater, Mars," Lunar and Planetary Science Conference, 48 (2017).

167 PLUNGING HUNDREDS OF METERS Emily Lakdawalla, "We're Going to Jezero!" The Planetary Society (Nov. 20, 2018).

167 BEVY OF SPACECRAFT If all goes well, also launching to Mars in 2020 will be the *Kazachok* lander, built by Roscosmos, the Russian space agency. It will collect images and monitor the weather at Oxia Planum, as well as deliver the European Space Agency's *Rosalind Franklin* rover to the surface to study rocks and soils. The rover will sink a drill deeper than ever before under the surface of Mars. One of its most exciting instruments, the Mars Organic Molecule Analyzer, will have the ability to detect a wide range of organics in the samples it collects. The Chinese also plan to launch a mission on one of their enormous *Long March 5* rockets. Their first Mars attempt, a joint mission with the Russians, sadly failed after lift-off in 2011, but in 2020, the China National Space Administration will go it alone with its *HX-1* mission. According to news sources, the orbiter will carry cameras, radar, spectrometers, neutral and energetic particle analyzers, and a magnetometer, and will be joined by a solar-powered rover (capitalizing on the success of China's *Yutu-2* rover, which explored the moon's far side in 2019). A spacecraft called *Hope* from the United Arab Emirates also plans to launch in 2020, on a Japanese rocket, with the goal of mapping Martian weather from orbit and studying climate dynamics and atmospheric escape. It will be the Arab world's first attempt at a mission to another planet, and a hugely impressive feat for a country that didn't even have a space agency until 2014. All these probes will join NASA's ongoing *InSight* mission, a geophysical lander that arrived in 2018 to study seismic activity and internal heat flow, as well as NASA's *Mars Odyssey* mission, which found water ice in the subsurface when it arrived in the early 2000s and now serves as a communications relay. Also looping the Martian skies are the European Space Agency's *Mars Express* mission; NASA's *Mars Reconnaissance Orbiter;* NASA's *MAVEN* mission; the *ExoMars Trace Gas Orbiter,* a joint effort between the European Space Agency and Roscosmos; and *Mangalyaan,* also known as the *Mars Orbital Mission,* the Indian Space Agency's thrilling first mission, which arrived in 2014 for less than the cost of making the movie *Gravity*. For more information, see: Emily Lakdawalla, "Similarities and differences in the landing sites of ESA's and NASA's 2020 Mars rovers," *Nature Astronomy,* 3 (2019), p. 190; Mike Wall, "4 Mars Missions Are One Year Away from Launching to the Red Planet in July 2020," *Space.com,* July 25, 2019; Andrew

Jones, "China's first Mars spacecraft undergoing integration for 2020 launch," SpaceNews, May 29, 2019; Sam Lemonick, "3 rovers will head to Mars in 2020. Here's what you need to know about their chemical missions." *C&EN,* 97, no. 29, July 21, 2019; Eshan Masood, "UAE Mars probe will be Arab world first," *Nature News,* July 31, 2019.

168 ALL BUT A FEW PATCHES Martin J. Van Kranendonk, Vickie Bennett, and Elis Hoffmann, eds., *Earth's Oldest Rocks* (Amsterdam: Elsevier, 2018).

168 POLAR CAPS WAX "The Changing Ice Caps of Mars," NASA Science Mars Exploration Program (Nov. 27, 2018).

168 SPIN AXIS ARCS Kevin W. Lewis, Oded Aharonson, John P. Grotzinger, Randolph L. Kirk, Alfred S. McEwen, and Terry-Ann Suer, "Quasi-Periodic Bedding in the Sedimentary Rock Record of Mars," *Science,* 322, no. 5,907 (2008), pp. 1,532–1,535; J. Taylor Perron and Peter Huybers, "Is There an Orbital Signal in the Polar Layered Deposits on Mars?" *Geology,* 37, no. 2 (2009), pp. 155–158.

168 FROM GEOTHERMAL FIELDS Armen Y. Mulkidjanian, et al., "Origin of First Cells at Terrestrial, Anoxic Geothermal Fields," *PNAS,* 109, no. 14 (2012), pp. E821–E830.

168 REPEATED CYCLES Jay G. Forsythe, Sheng-Sheng Yu, Irena Mamajanov, Martha A. Grover, Ramanarayanan Krishnamurthy, Facundo M. Fernández, and Nicholas V. Hud, "Ester-Mediated Amide Bond Formation Driven by Wet–Dry Cycles: A Possible Path to Polypeptides on the Prebiotic Earth," *Angewandte Chemie International Edition,* 54, no. 34 (2015), pp. 9,871–9,875.

168 ROVER'S CHASSIS The rover will also be delivered by a duplicate SkyCrane entry, descent, and landing system. Eric Hand, "NASA's Mars 2020 Rover to Feature Lean, Nimble Science Payload," *Science* (July 31, 2014).

168 SMALL HELICOPTER The helicopter drone weighs less than two kilograms and is just over a meter wide. It is designed for ultra-low density atmospheric flight, and if it succeeds, it will be the first aircraft flown on another body in the solar system. The helicopter features twin rotors and a solar array and will attempt five flights of up to 90 seconds above Mars's surface. Meant as a test craft, the rotorcraft carries no scientific equipment apart from its cameras, which will return images of Jezero Crater from above. See: D. C. Agle and Alana Johnson. "NASA's Mars Helicopter Attached to Mars 2020 Rover." NASA, 28 Aug. 2019 and Preston Lerner, "A Helicopter Dreams of Mars." *Air & Space Magazine,* April 2019.

168 JUST AS HEAVY The arm's instrument-laden turret weighs forty kilograms. Michelle Lou, "Watch the Arm of NASA's Mars 2020 Rover Perform a Bicep Curl," CNN (July 30, 2019).

169 DRILL SEVERAL SAMPLES The rover will carry forty-two sample tubes, including five engineering spares, but it is designed to acquire only twenty samples over the first 1.5 Mars years. The plan is for the rest to be acquired during the extended mission. Ken Farley, "Mars 2020 Mission," Fourth Landing Site Workshop for the Mars 2020 Rover Mission (Glendale, Calif., Oct.16, 2019).

169 SIZE OF A PENLIGHT "Robotic Arm," NASA *Mars 2020* Mission.

169 CACHE WILL REMAIN THERE The current strategy for returning the samples involves a fetch rover that will deliver samples to a Mars ascent vehicle, which will help prove technologies required for human exploration. At present, it is envisioned as an international endeavor, with the main mission hardware provided by NASA and the European Space Agency. "Sample Handling," NASA *Mars 2020* mission; Justin Cowart, "NASA, ESA Officials Outline Latest Mars Sample Return Plans," The Planetary Society (August 13, 2019).

169 AND BROUGHT HOME Even though a great deal of material has already been naturally transferred from Mars to Earth, and even though the deliberate goal of NASA's Mars Sample Return effort is to return ancient rocks with evidence of ancient life—not any living organisms—the potential for extant life makes it imperative that NASA and its international partners take extreme caution in handling any samples returned to Earth. NASA's Office of Planetary Protection, in collaboration with the Committee on Space Research (COSPAR), an international scientific body, has implemented strict policies to protect Earth from possible life-forms returned from space. Several sample quarantine scenarios are being explored, including the use of specially built Biosafety Level-4 containment facilities (BSL-4, those used for fatal diseases with no known vaccines or treatments) as well as possibilities in orbit. See: Bergit Uhran, Catharine Conley, and J. Andy Spry, "Updating Planetary Protection Considerations and Policies for Mars Sample Return," *Space Policy* (2019); Sarah Knapton, "Martian Rocks Could Be Quarantined on Moon Before Travelling to Earth," *The Telegraph* (London: Dec. 5, 2019); Yoseph Bar-Cohen, Mircea Badescu, Stewart Sherrit, Xiaoqi Bao, Hyeong Jae Lee, Erik Bombela, and Sukhwinder Sandhu, "Sample Containerization and Planetary Protection Using Brazing for Breaking the Chain of Contact to Mars," *Behavior and Mechanics of Multifunctional Materials XIII*, vol. 10968, International Society for Optics and Photonics (2019), p. 1,096,802.

169 THE PRECISE AGE Melanie Barboni, Patrick Boehnke, Brenhin Keller, Issaku E. Kohl, Blair Schoene, Edward D. Young, and Kevin D. McKeegan, "Early Formation of the Moon 4.51 Billion Years Ago," *Science Advances,* 3, no. 1 (2017), p. e1602365.

169 AN INDELIBLE RECORD Prabal Saxena, Rosemary M. Killen, Vladimir Airapetian, Noah E. Petro, Natalie M. Curran, and Avi M. Mandell, "Was the Sun a Slow Rotator? Sodium and Potassium Constraints from the Lunar Regolith," *The Astrophysical Journal Letters,* 876, no. 1 (2019), p. L16.

169 A MILLION PASSENGERS, SENT ON Sarah Knapton, "Elon Musk: We'll Create a City on Mars with a Million Inhabitants," *The Telegraph* (London: June 21, 2017).

169 HOME TO A RELICT RIVER DELTA Timothy A. Goudge, Ralph E. Milliken, James W. Head, John F. Mustard, and Caleb I. Fassett, "Sedimentological Evidence for a Deltaic Origin of the Western Fan Deposit in Jezero Crater, Mars, and Implications for Future Exploration," *Earth and Planetary Science Letters,* 458 (2017), pp. 357–365; Timothy A. Goudge, David Mohrig, Benjamin T. Cardenas, Cory M. Hughes, and Caleb I. Fassett, "Stratigraphy and Paleohydrology of Delta Channel Deposits, Jezero Crater, Mars," *Icarus,* 301 (2018), pp. 58–75.

169 MOST LIKELY TO TRAP THINGS Jack D. Farmer and David J. Des Marais, "Exploring for a Record of Ancient Martian Life," *Journal of Geophysical Research Planets,* 104, no. E11 (1999), pp. 26,977–26,995.

169 BIND AND BURY ORGANICS Bethany L. Ehlmann, et al., "Clay Minerals in Delta Deposits and Organic Preservation Potential on Mars," *Nature Geoscience,* 1, no. 6 (2008), p. 355.

170 ONE OF THE ROVER'S MAIN TARGETS Sanjeev Gupta and Briony Horgan, "Mars 2020 Science Team Assessment of Jezero Crater," Fourth Landing Site Workshop for the Mars 2020 Rover Mission (Glendale, Calif.: Oct. 17, 2019); Kennda Lynch, James Wray, Kevin Rey, and Robin Bond, "Habitability and Preservation Potential of the Bottomset Deposits in Jezero Crater," Fourth Landing Site Workshop for the Mars 2020 Rover Mission (Glendale, Calif.: Oct. 17, 2019); Ken Farley, Katie Stack-Morgan, and Ken Williford, "Jezero-Midway Interellipse Traverse Mission Concept," Fourth Landing Site Workshop for the Mars 2020 Rover Mission (Glendale, Calif.: Oct. 18, 2019).

170 RICH BOTTOMSET BEDS Hydrated silica, a mineral that's terrific for preserving microfossils and other traces of life, has also been detected near the edge of the delta in what may be its bottomset layer, adding to excitement of exploring this locality. Other exciting targets within Jezero include a bathtub ring of carbonates that may represent an ancient lakeshore. J. D. Tarnas, J. F. Mustard, Honglei Lin, T. A. Goudge, E. S. Amador, M. S. Bramble, C. H. Kremer, X. Zhang, Y. Itoh, and M. Parente, "Orbital identification of hydrated silica in Jezero crater, Mars," *Geophysical Research Letters* (2019); Briony H.N. Horgan, Ryan B. Anderson, Gilles Dromart, Elena S. Amador, and Melissa S. Rice, "The mineral diversity of Jezero crater: Evidence for possible lacustrine carbonates on Mars," *Icarus* (2019): 113526.

170 TRIANGLE SHAPE Donald Prothero and Fred Schwab, *Sedimentary Geology: An Introduction to Sedimentary Rocks and Stratigraphy (Second Edition)* (New York: W. H. Freeman and Co., 2004).

170 "ERASED FROM TIME" Haleh Ardebili, "No. 2822: Herodotus," Engines of Our Ingenuity (Aug. 21, 2012).

170 "AUTOPSY" Herodotus, *Histories*, trans. George Rawlinson (The Internet Classics Archive), II.

170 THINGS HE NOTICED Herodotus, *An Account of Egypt*, trans. G. C. Macaulay (Project Gutenberg, 2006).

170 "SILTING FORWARD OF THE LAND" Ibid.

170 *AKHET* Nigel C. Strudwick, *Texts from the Pyramid Age* (Atlanta: Society of Biblical Literature, 2005), p. 87.

170 ONLY CONSONANTS REMAINING Ibid.

170 SHAPED THE WET CLAY "Egyptian Pottery," Ceramics and Pottery Arts and Resources (July 30, 2009).

171 DECORATED THE JARS Ibid.

171 PATTERNS WOULD REMAIN Ibid.

171 THE PELUSIAC BRANCH W.W. How, *A Commentary on Herodotus* (Project Gutenberg, 2008).

171 HIGH AS FIVE METERS Joshua J. Mark, "Egyptian Papyrus," *Ancient History Encyclopedia* (Nov. 8, 2016).

171 WORLD WAS CREATED Janice Kamrin, "Papyrus in Ancient Egypt," The Metropolitan Museum of Art (March 2015).

171 "GIFT OF THE RIVER" J. Gwyn Griffiths, "Hecataeus and Herodotus on 'A Gift of the River,'" *Journal of Near Eastern Studies*, 25, no. 1 (1966), pp. 57–61.

171 RED-HOT OVENS Herodotus, *An Account of Egypt*, trans. G. C. Macaulay.

171 MADE GARLANDS From Theophrastus, *Historia Plantarum* (IV, 10), as quoted in "Papyrus," *Encyclopedia Britannica* (1911).

171 HERODOTUS'S WRITINGS FOUND Lionel Casson, "The Library of Alexandria," *Libraries in the Ancient World* (New Haven: Yale University Press, 2002), p. 43.

171 GREAT LIBRARY OF ALEXANDRIA Roy MacLeod, ed., *The Library of Alexandria: Centre of Learning in the Ancient World* (London: I. B. Tauris, 2004).

171 UNBROKEN SCHOLARSHIP Roy MacLeod, ed. *The Library of Alexandria: Centre of Learning in the Ancient World* (London: I. B. Tauris, 2004).

171 GREAT LIGHTHOUSE Judith McKenzie and Peter Roger Stuart Moorey, *The Architecture of Alexandria and Egypt, c. 300 BC to AD 700*, vol. 63 (New Haven: Yale University Press, 2007); James Crawford, *Fallen Glory: The Lives and Deaths of History's Greatest Buildings* (New York: Picador, 2017).

171 COPIED BY SCRIBES Claudii Galeni, *Opera Omnia*, vol. 17.1, ed. D. Carolus Gottlob Kühn (Leipzig, Prostat in Officina Libraria Car. Cnoblochii, 1828), p. 605.

171 COLLECTION GREW "When Libraries Were on a Roll," *The Telegraph* (London: May 19, 2001).

171 THE PERSIANS, THE BABYLONIANS Thomas Greenwood, "Euclid and Aristotle," *The Thomist: A Speculative Quarterly Review*, 15, no. 3 (1952), pp. 374–403.

171 *EUCLID'S ELEMENTS* Euclid, *Euclid's Elements: All Thirteen Books Complete in One Volume: The Thomas L. Heath Translation* (Santa Fe, N.M.: Green Lion Press, 2002).

173 PERFECT NUMBERS A perfect number is any positive integer that is equal to the sum of its proper divisors; for example, 6 is the sum of 1, 2, and 3 and 28 is the sum of 1, 2, 4, 7, and 14.

173 KANTIAN IDEAL Immanuel Kant, *Critique of Pure Reason*, trans. and ed. by Paul Guyer and Allen W. Wood (Cambridge University Press: 1998).

174 CARL FRIEDRICH GAUSS Gauss, who had an avid interest in extraterrestrials, is also the German rumored to have proposed the Pythagorean right triangle extraterrestrial-signaling scheme; whether or not the idea originated with him is not certain. Michael J. Crowe, *The Extraterrestrial Life Debate, 1750–1900* (Mineola, N.Y.: Dover Publications, 2011), p. 205.

174 THE PARALLEL POSTULATE Euclid's fifth postulate is: "If a straight line intersects two other straight lines, and so makes the two interior angles on one side of it together less than two right angles, then the other straight lines will meet at a point if extended far enough on the side on which the angles are less than two right angles." Euclid, *Euclid's Elements: All Thirteen Books Complete in One Volume: The Thomas L. Heath Translation.*

174 AVOIDED USING IT J. J. O'Connor and E. F. Robertson, "Non-Euclidean Geometry," JOC/EFR (Feb. 1996).

174 KEPT HIS DOUBTS A SECRET Ibid.

174 THEORY OF GENERAL RELATIVITY Albert Einstein, "Der Feldgleichungen der Gravitation," *Königlich Preußische Akademie der Wissenschaften* (1915), pp. 844–847.

175 PHOTOSYNTHESIS EVOLVED LATE Joseph R. Michalski, Tullis C. Onstott, Stephen J. Mojzsis, John Mustard, Queenie H. S. Chan, Paul B. Niles, and Sarah Stewart Johnson, "The Martian Subsurface as a Potential Window into the Origin of Life," *Nature Geoscience,* 11, no. 1 (2018), p. 21; Tanai Cardona, James W. Murray, and A. William Rutherford, "Origin and Evolution of Water Oxidation before the Last Common Ancestor of the Cyanobacteria," *Molecular Biology and Evolution,* 32, no. 5 (2015), pp. 1,310–1,328; Akiko Tomitani, et al., "The Evolutionary Diversification of Cyanobacteria: Molecular-Phylogenetic and Paleontological Perspectives," *PNAS,* 103, no. 14 (2006), pp. 5,442–5,447.

175 THE VAST MAJORITY OF MICROBES Yinon M. Bar-On, Rob Phillips, and Ron Milo, "The Biomass Distribution on Earth," *Proceedings of the National Academy of Sciences,* 115, no. 25 (2018), pp. 6,506–6,511.

175 DEEPEST SUBTERRANEAN MINES B. Lollar Sherwood, Georges Lacrampe-Couloume, G. F. Slater, J. Ward, Duane P. Moser, T. M. Gihring, L-H. Lin, and Tullis C. Onstott, "Unravelling Abiogenic and Biogenic Sources of Methane in the Earth's Deep Subsurface," *Chemical Geology,* 226, no. 3–4 (2006), pp. 328–339; Katrina J. Edwards, Keir Becker, and Frederick Colwell, "The Deep, Dark Energy Biosphere: Intraterrestrial Life on Earth," *Annual Review of Earth and Planetary Sciences,* 40 (2012), pp. 551–568; Cara Magnabosco, Kathleen Ryan, Maggie C. Y. Lau, Olukayode Kuloyo, Barbara Sherwood Lollar, Thomas L. Kieft, Esta Van HeerDen, and Tullis C. Onstott, "A Metagenomic Window into Carbon Metabolism at 3 km Depth in Precambrian Continental Crust," *The ISME Journal,* 10, no. 3 (2016), p. 730.

175 MORE OF THOSE LITTLE HOMES Joseph R. Michalski, et al.,"The Martian Subsurface as a Potential Window into the Origin of Life," *Nature Geoscience.*

175 STARTING TO UNDERSTAND T. C. Onstott, B. L. Ehlmann, H. Sapers, M. Coleman, M. Ivarsson, J. J. Marlow, A. Neubeck, and P. Niles, "Paleo-Rock-Hosted Life on Earth and the Search on Mars: A Review and Strategy for Exploration," *Astrobiology* (2019).

176 SITE CALLED MIDWAY Paul Voosen, "NASA's Next Mars Rover Aims to Explore Two Promising Sites." *Science* 362, no. 6411 (2018), pp. 139–140. Midway's strongest advocate in the landing site selection process was trailblazing Caltech professor Bethany Ehlmann, who has published extensively on Northeast Syrtis and has spearheaded a charge to explore ancient subsurface terrains.

176 RIDGES AND MESAS J. Mustard, et al., "Mars 2020 Candidate Landing Site Data Sheet: NE Syrtis," NASA JPL.

176 "LIFE AS WE DON'T KNOW IT" This research area is focus of the Laboratory for Agnostic Biosignatures (LAB) Project, for which I serve as Principal Investigator. With support from NASA's Astrobiology Program, LAB aims to develop life detection methods that identify unknowable, unfamiliar features and chemistries that may represent processes of life as-yet unrecognized. The LAB team includes biologists, chemists, computer scientists, mathematicians, and instrument engineers. Building on seminal work in the astrobiology community

(i.e., P. G. Conrad and K. H. Nealson, "A Non-Earthcentric Approach to Life Detection," *Astrobiology,* 1, no. 1 (2001), pp. 15–24), we are also designing tools for detecting these signatures and strategies for interpreting them (www.agnosticbiosignatures.com).

176 CHEMICAL COMPLEXITY S. M. Marshall, A.R.G. Murray, and L. Cronin, "A Probabilistic Framework for Identifying Biosignatures Using Pathway Complexity," *Philosophical Transactions of the Royal Society A: Mathematical, Physical and Engineering Sciences,* 375, no. 2109 (2017), p. 20,160,342; S. S. Johnson, E. V. Anslyn, H. V. Graham, P. R. Mahaffy, and A. D. Ellington, "Fingerprinting Non-Terran Biosignatures," *Astrobiology,* 18, no. 7 (2018), pp. 915–922.

177 DIFFERENT MOLECULAR FOUNDATION For instance, life might have evolved to use cosmic rays as a source of energy, just as we use visible wavelengths of light here on Earth, or to use sulfur, silicon, or ammonia in lieu of carbon as the building block of life. J. E. Lovelock, "A Physical Basis for Life Detection Experiments," *Nature,* 207, no. 997 (1965), pp. 568–570; P. H. Rampelotto. "The search for life on other planets: Sulfur-based, silicon-based, ammonia-based life," *Journal of Cosmology,* 5 (2010), pp. 818–827.

177 A THING "UNCUTTABLE" Some later scholars, it should be noted, have interpreted Aristotle's idea in *Physics* that there is a smallest size of material substrate on which it is possible for the form of a given natural tissue to occur (e.g., bone or blood) as evidence that Aristotle believed in the existence of minimal physical parts. See: Sylvia Berryman, "Ancient Atomism," in *The Stanford Encyclopedia of Philosophy*, Edward N. Zalta, ed. (Winter 2016 edition); "Scholastic Philosophy and Renaissance Magic in the De Vita of Marsilio Ficino," *Renaissance Quarterly,* 37 (1984), pp. 523–554; Ruth Glasner, "Ibn Rushd's Theory of Minima Naturalia," *Arabic Sciences and Philosophy,* 11 (2001), pp. 9–26; John E. Murdoch, "The Medieval and Renaissance Tradition of *Minima Naturalia,*" in Christoph Lüthy, John E. Murdoch and William R. Newman, eds., *Late Medieval and Early Modern Corpuscular Matter Theories* (Leiden: Brill, 2001, pp. 91–132).

177 NOT BUILT FROM INANIMATE PARTS For a fascinating discussion of worldviews, see the work of philosopher Lucas Mix, who discusses how Lucretius, an Epicurean atomist, viewed the world as fundamentally dead whereas Aristotle viewed the world as fundamentally alive, and "by and large Aristotle won for two thousand years." Lucas John Mix, "The Meaning of 'Life': Astrobiology and Philosophy," University of Washington Seminar Series, NASA Astrobiology Institute Virtual Planetary Lab, 12 May 2012, and Lucas John Mix, *Life in Space: Astrobiology for Everyone* (Cambridge: Harvard University Press, 2009), pp. 246–247; see also: Mohan Matthen and R. James Hankinson, "Aristotle's Universe: Its Form and Matter," *Synthese,* 96, no. 3 (1993), pp. 417–435.

177 THE HEAVENS Aristotle, "Book IV," *On the Heavens,* trans. W. K. C. Guthrie (Cambridge: Harvard University Press, 1939).

177 MANY OF THEM WERE WRONG Alan Lightman, *Searching for Stars on an Island in Maine* (New York: Pantheon Books, 2018), p. 145; Armand Marie Leroi, "6 Things Aristotle Got Wrong," *Huffington Post* (Dec. 2, 2014).

178 PARADIGM HAD SHIFTED In the 1960s, Thomas Kuhn proposed the idea of paradigm shifts in science: the idea that a dominant mode of scientific thought

slowly accumulates errors, then dominant thinking radically shifts to a new paradigm. T. S. Kuhn, *The Structure of Scientific Revolutions* (Chicago and London: University of Chicago Press, 1962).

178 MECHANICAL, CLOCK-LIKE UNIVERSE "Newtonian Cosmology and Religion," *Cosmic Journey: A History of Scientific Cosmology* (American Institute of Physics).

178 "IF THE DOORS OF PERCEPTION" William Blake, "The Marriage of Heaven and Hell," in *The Poetical Works*, ed. John Sampson (Oxford University Press, 1908; Bartleby.com, 2011), lines 115–116. Similarly, Einstein once said, "We are in the position of a little child entering a huge library filled with books in many languages. The child knows that someone must have written those books. It does not know how. It does not understand the languages in which they are written. The child dimly suspects a mysterious order in the arrangements of the books but doesn't know what it is. We see the universe marvelously arranged and obeying certain laws but only dimly understand those laws." Quoted in Alan Lightman, *Searching for Stars on an Island in Maine* (New York: Pantheon Books, 2018), pp. 115–116.

178 MEDIEVAL EARTHQUAKES Allison McNearney, "The Buried Secrets of the World's Very First Lighthouse," *The Daily Beast* (Oct. 21, 2017).

178 THE MECHANICAL BIRDS James Crawford, "The Life and Death of the Library of Alexandria," *LitHub* (March 13, 2017).

179 PARTICLES AND FORCES For a beautiful meditation on a scientist's belief in the material world, see: Alan Lightman, *Searching for Stars on an Island in Maine* (New York: Pantheon Books, 2018).

180 RIVETING LETTERS *Papers of William Henry Pickering, 1870–1907* (Harvard University Archives, HUG 1691).

180 INTERLEAVED ARE HUNDREDS Ibid.

180 SILENT-FILM FOOTAGE "Hallo, Mars? Mr. Marconi and his new aerial on the radial yacht 'Elettra,' with which he hopes to call up Mars," British Pathé.

180 A PICTURE FROM AROUND 1910 David Peck Todd papers, 1862–1939 (inclusive), Yale University Manuscripts & Archives.

180 WORLD IS GETTING QUIETER Duncan H. Forgan and Robert C. Nichol, "A Failure of Serendipity: The Square Kilometre Array Will Struggle to Eavesdrop on Human-like Extraterrestrial Intelligence," *International Journal of Astrobiology*, 10, no. 2 (2011), pp. 77–81.

180 WITHIN THEIR OWN EYES Lowell observed linear features not only on Mars but also Mercury ("cracks"), the moons of Jupiter ("lineaments"), and Venus ("spokes"). The similarities between the "spokes" he charted on a map of the surface of Venus at the turn of the twentieth century and a modern image of blood vessels diverging from the optic cup are uncanny; see: W. Sheehan and T. Dobbins, "The Spokes of Venus: An Illusion Explained," *Journal for the History of Astronomy*, vol. 34, part 1, no. 114 (2003), p. 61.

180 "LITTLE GOSSAMER FILAMENTS" Percival Lowell, *Mars as the Abode of Life* (New York: The Macmillan Company, 1908), p. 146.

180 "UNDESCRIBED, IMPERFECT YEASTS" Helen S. Vishniac and Walter P. Hempfling, "*Cryptococcus Vishniacii* sp. nov., an Antarctic Yeast," *International Journal of Systematic and Evolutionary Microbiology,* 29, no. 2 (1979), pp. 153–158.

181 TENDED TO THE SLIDES Ephraim Vishniac, phone interview by Sarah Johnson (Sept. 8, 2017).

181 THE EDITION Euclid, *Euclid's Elements: With Notes, an Appendix, and Exercises by Issac Todhunter* (independently published, 2017). The painting is Caspar David Friedrich, *Wanderer above the Sea of Fog,* 1818, oil on canvas (2007).

181 "TRACES OF HUMAN EVENTS" Herodotus, quoted in Thomas Harrison, "Review: Herodotus," *The Classical Review,* 49, no. 1 (1999), p. 16.

181 AS MANY AS FORTY BILLION Erik A. Petigura, Andrew W. Howard, and Geoffrey W. Marcy, "Prevalence of Earth-Size Planets Orbiting Sun-like Stars," *Proceedings of the National Academy of Sciences,* 110, no. 48 (2013), pp. 19,273–19,278; Melissa Block, "Study Says 40 Billion Planets in Our Galaxy Could Support Life," NPR, *All Things Considered* (Nov. 5, 2013).

181 SPEED LIMIT For an illuminating discussion of limits in science, from the speed of light as the speed limit of the universe to the limits Heisenberg uncovered about what can be measured with certainty, see: John D. Barrow, *Impossibility* (Oxford University Press, 1998).

182 ACROSS MERIDIANI PLANUM The sunset images were collected on November 4 and 5, 2010. Guy Webster, "Martian Dust Devil Whirls Into Opportunity's View," NASA (July 28, 2010); Webster, "Mars Movie: I'm Dreaming of a Blue Sunset," NASA (December 22, 2010).

182 THE MICROPHYSICS OF THE SYSTEM On Earth, blue light is scattered more efficiently than red light, which has a longer wavelength. This phenomenon, Rayleigh scattering, results from light scattering in all directions off of objects that are much smaller than the wavelength of light, in this case the very small molecules of air in our atmosphere. This is why our sky is blue. But by the time incident light from the sun has traveled through a great deal of atmosphere (e.g., at sunset, when the sun is low on the horizon), the blue light has been mostly scattered away. The result is that mostly red and yellow photons hit our eyes. On Mars, there is very little air, so Rayleigh scattering is less prominent. Most of the light is scattered by dust particles, which tend to be similar in size or larger than the wavelengths of incident light. As the light is scattered, the blue wavelengths are deflected less than red wavelengths, and sunsets on Mars appear blue. Recent observations from the Compact Reconnaissance Imaging Spectrometer aboard the Mars Reconnaissance Orbiter have led to important advances in characterizing the scattering properties of Martian aerosols; see: M. J. Wolff, M. D. Smith, R. T. Clancy, R. Arvidson, M. Kahre, F. Seelos, S. Murchie, and Hannu Savijärvi, "Wavelength dependence of dust aerosol single scattering albedo as observed by the Compact Reconnaissance Imaging Spectrometer," *Journal of Geophysical Research: Planets* 114, no. E2 (2009).

Index

ABOUT THE AUTHOR

SARAH STEWART JOHNSON is an assistant professor of planetary science at Georgetown University. A former Rhodes Scholar and White House Fellow, she received her PhD from MIT and has worked on NASA's *Spirit, Opportunity,* and *Curiosity* rovers. She is also a visiting scientist with the Planetary Environments Lab at NASA's Goddard Space Flight Center.